This book is written in honor of the Kaua'i Chamber of Commerce and the work of its officers and members over the past nine decades. The Chamber continues today in its long tradition as a leading force for positive change on Kaua'i. The book also honors and aims at perpetuating the story behind the work and commerce of the people of Kaua'i.

The support of local businesses made possible the publication of *Kaua'i: The Garden Island, A Pictorial History of the Commerce and Work of the People.* The Kaua'i Chamber of Commerce is pleased to acknowledge these sponsors and say mahalo for their support:

Alexander & Baldwin, Aston Hotels & Resorts, Gay & Robinson, Jack Harter Helicopters, Kauai Electric, Kauai Power Partners, King Auto Center, Niu Pia Farms, Pahio Companies.

Sponsored by:

Alexander & Baldwin
Aston Hotels & Resorts
Gay & Robinson
Jack Harter Helicopters
Kauai Electric
Kauai Power Partners
King Auto Center
Niu Pia Farms
Pahio Companies
Kaua'i Chamber of Commerce

Alekoko fish pond at sunset. The ancient rock walled fish catchment is commonly known as Menehune Fishpond.

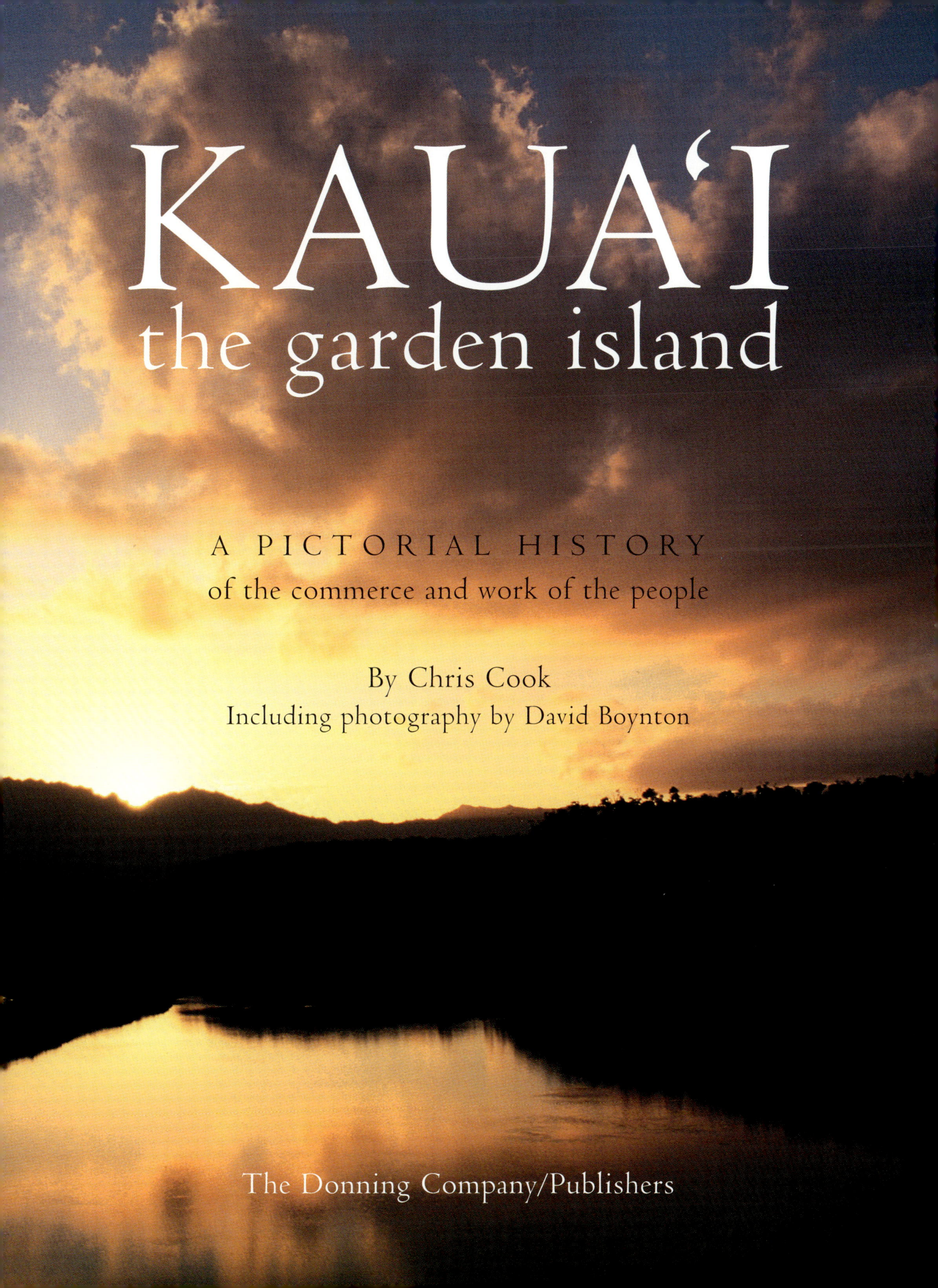

KAUA'I
the garden island

A PICTORIAL HISTORY
of the commerce and work of the people

By Chris Cook
Including photography by David Boynton

The Donning Company/Publishers

For the ancient ones of Kaua'i, the Western pioneers, those gathered from across the globe to work in the fields, and the newcomers, all.

The Donning Company/Publishers
184 Business Park Drive, Suite 106
Virginia Beach, VA 23462

Steve Mull, General Manager
Dawn V. Kofroth, Assistant General Manager
Barbara A. Bolton, Project Director
Diana Glynn, Project Research Coordinator
Shannon H. Garza, Associate Editor
Julia Kilmer-Buitrago, Associate Editor
Kelly M. Perkoski, Senior Graphic Designer
John Harrell, Imaging Artist
Teri S. Arnold, Director of Marketing

Library of Congress Cataloging-in-Publication Data:

Cook, Chris, 1950–
Kaua'i, the Garden Island : a pictorial history of the commerce and work of the people / by Chris Cook ; including photography by David Boynton.
p. cm.
Includes bibliographical references and index.
ISBN 1-57864-072-5 (hardcover : alk. paper)
1. Kauai (Hawaii)—History. 2. Kauai (Hawaii)—History Pictorial works. 3. Kauai (Hawaii)—Economic conditions. 4. Kauai (Hawaii)-
-Economic conditions Pictorial works. I. Boynton, David. II. Title. III. Title: Kaua'i.
DU628.K3C66 1999
996.9'41—dc21
99-34463

CIP
Printed in the United States of America

table of contents

The filming of South Pacific at Makua Beach, Ha'ena in the summer of 1957 provided work and excitement for hundreds of Kaua'i residents.

GO FOR
BROKE

foreword

The Island of Kaua'i is very special to me. In September of 1899, my grandparents and my father, then an infant, arrived at Nawiliwili Harbor from faraway Japan. They came to work in the sugarcane fields of the McBryde Sugar Company. Once fearful of the future in an unknown place, they were soon assured of a good life for themselves and their children.

In my Washington, D.C., office, I have framed and hanging on my wall the original contract my grandfather signed with an "X" one hundred years ago. It serves as a constant reminder that through hard work, sacrifice, and determination almost anything is possible.

I was reminded of this age-old truism in the aftermath of Hurricane 'Iniki. In the midst of the devastation and despair were sparks of hope filled with the sheer determination that Kaua'i will rise again. It has taken hard work, and there is much more that must still be done. Together with the members of the Kaua'i Chamber of Commerce and the vigilant spirit of the people, we will be successful.

Building upon the strength of its people, Kaua'i has much to offer—more than its balmy weather, lush valleys, majestic mountains, and pristine oceans. Kaua'i is not only a place to visit, it is an excellent place to do business. Its business climate is friendly and open, and striving to keep up with changing times. The Kaua'i Chamber of Commerce has been a leader, supporting the old and the new, the farmer, the computer engineer, and the unique handcrafter. Together, they make up the vibrant economy and the people of Kaua'i.

Kaua'i: The Garden Island, A Pictorial History of the Commerce and Work of the People does a fine job of telling Kauai's story. The pictures capture the vitality and determination of the people to succeed and prosper. For me, it captures the resilience of the human spirit and the power of hope. I recommend it for your business and reading pleasure.

Congratulations to the Kaua'i Chamber of Commerce for a job well done, and for your service through the years. For all of you who have linked your lives and your fortunes to the Garden Island and its promising future, I wish you all the best. It is indeed a wise investment.

—Daniel K. Inouye
United States Senator

LIHUE
ICE
CO.

preface

The commerce and work of the people of Kaua'i have a special context—that of an isolated island over two thousand miles away from the nearest continental land mass. The joy of the rural lifestyle found on Kaua'i amidst spectacular scenic beauty is a trade off. The security of life in big-city Honolulu or on the mainland is beyond the horizon. Probably this struggle to survive touches one more on Kaua'i than most counties in the United States, especially in the years following the devastation of a hurricane, as the Island experienced following 'Iniki in 1992 and 'Iwa in 1982.

The story of Kaua'i illustrated and told in the following pages is one of soil and sweat, of waves of new ideas and technologies, of men and women from every corner of the globe gathered on the Garden Island of Hawai'i to work out their destiny in a very special location. It is a story of innovators, idealists, and individualists. It is a story filled with pathos—drama, tragedies, and triumphs, one of making sacrifices, sometimes making do with little. It is the story of a multitude of ethnic groups living and working together in the spirit of aloha that emanates from Kauai's host culture, the native Hawaiians. But the sacrifice is more than worth it to those who love life on Kaua'i.

Clockwise from top left:
Early hand-blown soda bottle with topper.

Workers hull Kaua'i coffee cherries probably grown at Hanakapi'ai by the Deverill family.

Kaula, Ni'ihau, Lehua on a crystal clear winter day.

Expert irrigation tunnel diggers marked their work with Japanese characters on a hillside above Waimea Valley.

acknowledgments

by Chris Cook

Understanding Kauai's past is a complex undertaking. It is a long voyage, and one that requires the gathering of information, and more important the understanding of how that information fits into the tapestry of time that shows the full picture of the past. Mahalo and my everlasting gratitude to those who have helped me move along on this voyage in ways both big and small.

Thanks to former Kaua'i Chamber of Commerce President Laurie Yoshida for considering me for this project, and to current President Mamo Cummings for helping me finish it, and to Chamber chairpersons Donna Apisa and Bob Mullins, to the Board of the Chamber and its members for their backing. To sponsors Jack & Bev Harter and Casey Riemer; Alexander & Baldwin; E. Alan Kennett at Gay & Robinson; Lynn McCrory and the management and staff at Pahio Resorts; Charlie King and his crew at King Auto; John Burns at Kauai Power Partners; Denny Polasky and Jenny Fujita at Kauai Electric; to Aston Hotels, especially my friend Lopaka Mansfield and others at Waimea Plantation Cottages; and to Niu Pia Farms. To David Boyton, my compatriot, benevolent critic and co-seeker of the past, Sue Boynton, and Betty Boynton. To Barbara Bolton, Julia Kilmer-Buitrago, and Kelly Perkoski of The Donning Company Publishers. To Nick Beck; Carlos Andrade; Alan Fayé; Steve Kyono; to Barnes Riznik and Marion Penhallow. To the memory of Edward Joesting, Catherine Stauder, Bob Gehran, and Ethel Damon. To Carol Lovell, Christine Fayé, and Margaret Lovett at the Kauai Museum; Carolyn Larson, Joe Ursprung, John Lydgate, Jay Furfaro, and Mary Requilman at the Kauai Historical Society; Cynthia Schur, Lynne Cosner, Rita De Silva, Sue Dixon, Dennis Fujimoto and all the cast and crew at *The Garden Island*; to Turk Tokita; Mayor Maryanne Kusaka & staff; Chris Deverill; Dave Walker; Raymond Massey and Fred Von Weigen; KTOH's Mike Ashman; Marilyn Wong; Warren Onishi; Julie Yukimura; the family of Harvey Tam; John Wehrheim and JoAnn Yukimura; Richard Sugiyama; Kauai Film Commissioner Judy Drosd; C. W. Spitz III, and the crew at OED; Howard Daniel, Yvette Ho, and Linda Howe at Alexander & Baldwin; Kikiaola Land Co.; the Cassell family of Waimea Valley; Dolly Strazer; Jill Landis; Pat Durkin; Tani Bova; Tony Sommer; Nick Galante and Vida Mossman at PMRF; Dave Alexander; John Kruse; Jeff White at the State Archives; Bob Kajiwara and staff at KCC; the librarians at Lihu'e, Kapa'a, Koloa, and Waimea; Will Mauck at Tin Can Mailman; Patty Ewing; Alan Drinan; Millie Wellington; Kalani Flores and the crew at Elderhostel; my Interpretive Guide students, Bill Blackburn, and Bobbie Franklin at KCC; Frances Frazier; Maile and Bob Semitekol; Freckles Smith; Sue Kanoho; Bill Gray; Nancy McMahon; and especially to my parents Lee and Sunny Cook for their support, and to my sons David Cook and Christian Cook, and my wife Evelyn Cook, for allowing me time for researching and writing.

CHAPTER I

The secrets of Kauai's past are slowly being uncovered by geologists, botanists, and archaeologists armed with modern high-tech tools.

However, the secret of its place name's source may never be known. It is known that Kaua'i and Ni'ihau are Hawaii's oldest main islands, about five million years old, according to geologists. Also in question is how many volcanoes formed the Islands. Geologists do agree that the windward face of Wai'ale'ale and the shape of the peak focus moist trade wind–borne clouds atop the 5,148-foot-high peak, making it one of the wettest spots on Earth.

A huge sea mount lies under Kaua'i and Ni'ihau, invisible to surface dwellers. The undersea volcanic mountain thrusts up from the sea floor eighteen thousand feet below the Pacific. Data gathered by NASA at their Koke'e tracking station shows that Kaua'i continues to move towards Japan at a rate of 3.4 inches per year as the huge tectonic plate it sits on creeps northwest.

ancient roots

Geologists say the wind, rain, and collapses of huge sections of the Island are the chief tools behind the fantastically eroded mountain peaks of Makaleha, Ha'upu, Kalalea, and above all Wai'ale'ale and Kawaikini. Millions of years of waves beating on the coral shelves provided the dozens of white sand beaches laced with green olivine, tiny particles of basalt, sea shell fragments, and bits of coralline algae.

The natural work of the Earth set the stage for the livelihood of Kauai's people, creating the bays of Hanalei and Nawiliwili; the flatlands for sugarcane growing up and down the west coast from Kekaha to Koloa, and along the eastern plain from Kipu to Kilauea; and the rivers and streams to water taro *lo'i* at Hanalei Valley, Hanapepe, and Waimea.

An early geological model shows Kauai's serrated mountains.

The first organic life came from the sea, beginning with the growth of coral reefs in secluded bays away from freshwater streams. Offshore pelagic game fish like the *ahi* and *aku* swam, and reef fish arrived, many cousins to those of the South Pacific. The ancient green sea turtle and the Hawaiian Monk Seal probably ruled as kings of this early world. Life on land began when the gray-black basalt rock oxidized, rusting into square miles of red soil. Plant seeds brought by the winds and sea currents, and perhaps migratory birds, arrived and grew into grasslands, rain forests, coastal ground covers, *loulu* palms, and other one-of-a-kind endemic plants still being discovered in secluded mountain tops and valleys.

Birds and hoary bats made their way to the Island to live and adapt in unique ways to the ecosystems that formed, ranging from arid, desert-like conditions on the

The stones of Pu'u-Ka-Pele overlooking Waimea Canyon are from the original lava flows that formed Kaua'i.

leeward sides, and to windward, verdant, lush rain forests.

To this pristine island came man. Exactly when remains a mystery, with arrival dates ranging from 250 A.D. to 800 A.D. for the arrival of canoes from the Marquesas and Tahiti in the South Pacific, and perhaps South America and the western Pacific.

A specific answer to the first arrival is being sought by paleontologist Dr. David Burney of Fordham University. At a dig along the sides of a sinkhole along the coast at Maha'ulepu, Burney and his team are finding bones of the *Moa Nalo*—an oversized, flightless duck-like bird—and carbon from fires lit by earlier voyagers.

Settlements were started in only a handful of seacoasts and valleys. As waves of voyagers arrived, the aboriginal settlers might have been driven into the interior forest, as is reflected in the legends of the *Menehune* translated by William Hyde Rice and other Kaua'i myth collectors.

The use of specialized stirrup and ring *poi* pounders only on Kaua'i, and the inclusion of Marquesan words in the Kaua'i dialect when Captain James Cook arrived in 1778, point to the eastern Polynesian island group as a likely homeland for settlers.

We do know these voyagers were brilliant craftsmen who used only Stone Age tools, had a diverse social structure and complex language.

They found a pristine island with many plants new to them, many birds along the coast and in the forests, and reefs rich in fish and shellfish. Few edible plants grew on Kaua'i before the voyagers arrived, and they carefully planted cuttings cared for during their voyage. Before the establishment of taro terrace systems, the growth of flocks of domestic animals and the figuring out of fishing patterns, their diet consisted of flightless birds, shellfish, reef fish, and other ocean life they ate raw, broiled,

At the turn of the century, native plants lined streams at Koke'e.

Waiehu, the traditional name of Wailua Falls, marks the convergence of older lava flows with more recent eruptions.

Kaua'i lei makers specialize in creating mokihana and maile leis.

steamed, and roasted in underground ovens.

Early Polynesian voyagers introduced over two dozen different plants essential to life. Along the shore they studied the rocks and crevices of the coast. During certain seasons a *kapu* was placed on the fishing grounds and enforced by a local *konohiki* or overseer. Fishermen found a less diverse marine environment than in the South Pacific, though the fringing reefs at 'Anini, Pila'a, and Ha'ena were similar to those of Tahiti. Bone carvers developed fish hooks specialized for Kaua'i conditions, and canoes were set up to troll offshore for *mahimahi*, *aku*, and *ahi*.

Coastal communities shared a *halau*, or a canoe house. There canoes were stored. They were carved with *adzes* from koa and other trees felled along the rim of Waimea Canyon and in nearby valleys. They became expert at building *hale* using *pili* grass and *lau hala* for roofing. Hard basalt stones shaped as *adzes* were imported from quarries on O'ahu and other islands and used to cut wood. Rubbing rock against rock, *poi* pounders, *kukui* nut lamps, and other stone implements were made. Daily work for women might include weaving baskets from *lau hala*, and pounding the bark of a *wauke* tree into cloth. Specialized crafts included feather and shell working, and tattooing.

By 1000 A.D. the roots of the Kaua'i society that met Cook were in place. As the population grew from the *'ohana* of early arrivals, political divisions set off in land sections were formed. Five *moku* divisions were made, the largest Kona, to the west, and Puna, on the east shore, plus Na Pali to the northwest, Ko'olau on the northeast corner, and Halelea'a along the North Shore from Ha'ena to Kalihiwai. Within the *moku*, *ahupua'a* developed, providing controlled access to resources from the reefs and seashore, across the plains on up to highland forests. About fifty-five

Center:
Ring and stirrup poi pounders were manufactured and used only on Kaua'i.

Right:
The moku, or major land sections of Kaua'i, were the traditional districts of the island into the late 1800s.

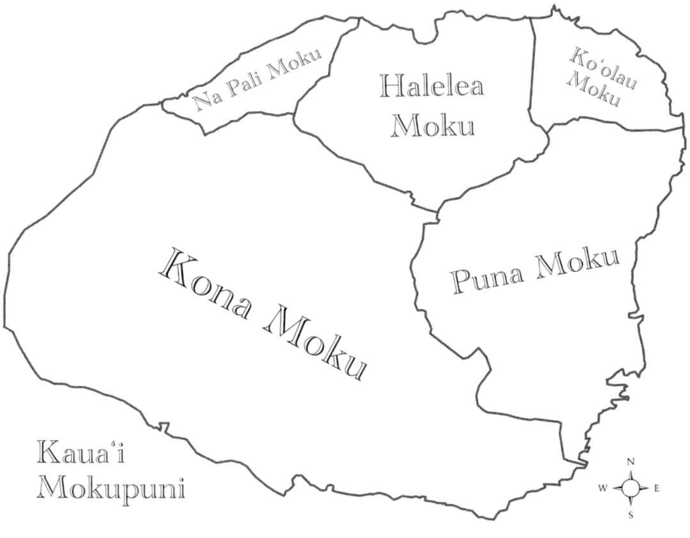

Above:
Legend says the dressed stones that carry the waters of Kiki-A-Ola were hewn by the Menehune.

Below:
Traditional hula practitioners use puili, or split bamboo rattles, in performing a traditional Kaua'i hula. Such practitioners performed in remote sections of Kaua'i throughout the 1800s and into the 1900s.

ahupua'a existed in the 1800s. Within *ahupua'a* smaller sections known as *'ili* and *kuleana* were formed.

Legendary rulers rose up in time. Among the most famous is Mo'ikeha, the voyager from Waipio Valley on the Big Island who married a daughter of Puna, a king of Kaua'i with a capital at Wailua. Up until the arrival of Captain Cook the leeward kingdom of Kaua'i experienced epochs of war and peace, of great advancements in irrigation and agriculture under benevolent rulers like Mano-ka-lani-po.

When Cook landed at Waimea in January 1788 Kamakahelei, the mother of Kaumuali'i, Kauai's last king, was queen. The queen's great-grandfather Kuali'i united Kaua'i, O'ahu, Moloka'i, and part of Maui. During her reign, Kaua'i and Ni'ihau stood alone as the leeward, separate kingdom of the Hawaiian chain.

Above:
The salt pans at Pu'olo Point at Hanapepe are the last major Hawaiian salt making site on Kaua'i.

Center:
The ancient taro lo'i in Hanapepe Valley once covered its floor. By 1900 many of the lo'i were planted in rice.

Below:
Until modern times, the entrance to Maninibolo, the "Dry Cave" at Ha'ena, was lined with a rock wall.

Above:
Some of the last Hawaiian hale on Kaua'i were located at Na Pali. The ridge line of the hale, built to keep rain out, was a distinctive feature of Kaua'i hale builders.

Below:
A lau hala weaver creates a mat at Wainiha.

CHAPTER 2

Captain Cook's arrival off Waimea on January 20, 1778 is pictured in this detail of a pencil drawing used by maritime artist Raymond Massey to paint the scene. Massey's work is released exclusively by Ship Store Galleries owner Fred Von Weigen of Wailua.

captain cook to annexation

Captain Cook

On January 18, 1778, Captain James Cook unexpectedly spotted the high mountains of an uncharted island. Wintry trade winds drove his ships HMS *Resolution* and HMS *Discovery* to the west, past the cloud-capped island. The next morning the ships pulled close to an island they would call Atooi, almost becalmed off Kawai Point and Kipukai on the southeast tip of Kaua'i.

The famed British navigator was amazed to find Polynesian traders aboard canoes that pulled out from shore to greet him, seeking pieces of iron in exchange for pigs and other provisions. Moving along the coast, heading northwest, Cook sought a river mouth from which to draw fresh water. On January 20 his ships anchored off Waimea, and on January 21 Cook went ashore. The party walked up Waimea Valley, and Cook was taken with the agricultural works of the people, passing the Menehune Ditch, a *heiau* and verdant fields.

In his journal he wrote: "Our journey to and from this morai, lay through the plantations. We observed most of the ground was perfectly flat, with ditches intersecting different parts, and roads that seemed to have been raised to some height by art. The intervening spaces . . . were planted with taro, which grew with great vigor. There were several spots where the cloth-mulberry was planted, in regular rows; this also grew vigorously. The coco-trees were in a less thriving condition, and were all low; but the plantain-trees made a pretty good appearance. . . . The greatest part of the village is near the beach, and consists of upward of sixty houses there, and we saw near forty more scattered about towards the *morai*. . . . we found a multitude collected at the beach, and a brisk trade for fowls, pigs, and vegetables, going on."

On February 2 the expedition sailed for the north to begin a search for a northwest passage from the Pacific to the Atlantic.

Cook's ships returned to provision following his death at Kealakekua on the Big Island in 1779, but no other European ships anchored off Kaua'i until a few years after the publication of Cook's journals in 1784 in London. The journals roughly mapped out Kaua'i and Ni'ihau and opened the door to Western exploration.

In 1786, HMS *King George* and HMS *Queen Charlotte*, under the command of Nathaniel Portlock and George Dixon, anchored off Waimea. Both captains had sailed with Captain Cook and were seeking a likely port for provisioning, repair, and rest stops for fur trades, sailing between Canton and the Pacific Northwest.

Young Kaumuali'i greeted the English explorer George Vancouver in March 1792. By this time three Western sailors lived at Waimea, Vancouver noted. On a return visit in 1793, Vancouver sailed off Wailua and Kapa'a and called the area "the most fertile and pleasant district of this island." On a third, and final, visit

A statue of Captain Cook first stood at Waimea High School in 1978. The Cook statue is a copy of one that stands in his home port of Whitby, England. The statue was moved to Hofgaard Park at Waimea in 1987.

made in 1794 Vancouver became the first westerner to map the north coast from Kilauea to Waimea, crossing Na Pali.

Kamehameha I

In 1795 Kamehameha brought all the Hawaiian Islands under his control, except for Kaua'i and Ni'ihau. The next spring Kaumuali'i became the king of Kaua'i and Ni'ihau as Kamehameha launched an invasion fleet from Waianae, O'ahu, intent on conquering the leeward islands. Lono-helemoa, a Hawaiian prophet, warned Kamehameha not to invade, but was rebuffed. Sailing at midnight, the fleet met strong winds and many canoes were overturned causing Kamehameha to suffer the greatest defeat of his life. Kaumuali'i fortified the island and armed men. In 1804 Kamehameha gathered another army, probably near Kualoa in windward O'ahu, to invade Kaua'i. His forces included seven thousand Hawaiians and fifty Europeans armed with muskets, a fleet of war canoes with cannons and swivel guns, and schooners. This time a typhoid-like disease struck the invaders, decimating Kamehameha's army and navy, and almost taking the king's life.

A chastened Kamehameha saw political means might be the way to conquer Kaua'i. He sent a messenger to Kaumuali'i and arranged a meeting in Honolulu in 1810. There, Kamehameha promised no bloodshed and continued control of his islands for his lifetime, if Kaumuali'i would come

The minting of a special U.S. half dollar and postage stamp marked the 150th anniversary of Cook's discovery of the Hawaiian Islands.

A native Hawaiian woman chants of Kauai's ancient past at the foot of a monument to Cook at Waimea during the 1928 observance of the 150th anniversary of the British explorer's arrival.

under his sovereignty. Kaumuali'i agreed, and Kamehameha told him to return and rule in peace. A tax, consisting of *kapa* cloth, mats, oranges, coconuts, calabashes, spears, hogs, and other articles, was agreed upon, and the ruler of Kaua'i returned home.

Furs and Sandalwood

The publication of Cook's journals in London in 1784 noted that Chinese merchants at Canton showed interest in furs from the Pacific Northwest. This simple statement sparked international interest in fur trading. In 1789 the fur trading ship *Columbia* became the first American vessel to anchor off Kaua'i. Inadvertently, the *Columbia* launched the sandalwood industry in Hawai'i when logs of the fragrant root wood cut on Kaua'i were burned as firewood.

The Chinese highly prized sandalwood, using the aromatic wood to make chests, boxes, in perfume and cosmetics, and as incense for their temples and religious ceremonies. American traders found the Chinese accepted few items in trade, mostly furs and Spanish Pieces of Eight.

The heart of the sandalwood tree provided the most aromatic wood and the trees were dug out of the ground, decimating forests of the slow-growing tree. The years 1811–1812 marked the beginning of a flourishing sandalwood industry.

During the War of 1812 American ship captains negotiated a contract with Kaumuali'i for a ten-year monopoly on sandalwood. The king received a quarter of the value of the wood carried to China. Trade picked up following the end of the war in 1815. The *maka'ainana*, or commoners, were ordered to leave their fishing grounds and taro *lo'i* to gather sandalwood. First, stands of sandalwood near the coast were worked, at Kalepa Mountain behind Hanama'ulu, in the hills behind Waimea and Wailua, and along the Ha'upu Range behind Koloa.

Sandalwood was measured by the Chinese weight called a *picul*, which equaled 133 1/3 pounds and was valued at ten to fifteen dollars. As many as five hundred canoes were used to bring off the wood set into a hole in the ground approximately the same length, width, and depth as a waiting ship's hold. Once accessible trees were stripped, the commoners were ordered into the inland mountains to cut sandalwood and carry the *piculs* on their backs for long distances. Waimea was the main port for shipping sandalwood, and some loads were carried there from Koloa and Wailua. Many collectors died of exposure in the cool forests.

A financial panic in 1819 cut the capital of the traders and made the trade in sandalwood more lucrative. By May 1821, Kaumuali'i went heavily into debt when he promised seventy-seven hundred *piculs* in exchange for a brig and its cargo of Western goods. The collecting peaked in 1821–1822. By then the abuse of the commoners by the *ali'i* created a break in the ancient mutually beneficial relationship. By the winter of 1822–1823 it was reported that the once

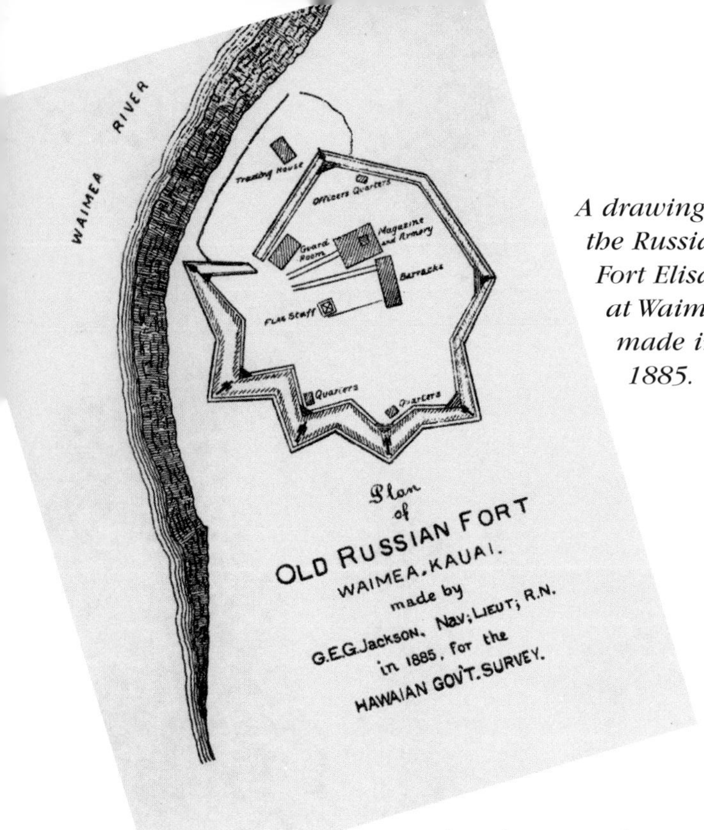

A drawing of the Russian Fort Elisabeth at Waimea made in 1885.

abundant supply on Kaua'i was nearly gone. Towards the end commoners were given a cut of the trade as an incentive.

The Russian Incident

Kaumuali'i had second thoughts about his pact with Kamehameha when a Russian expedition arrived under the leadership of a Dr. Scheffer, a Prussian doctor. The Russians directed Hawaiians who labored to construct a fort at Hipo, a coastal flatland on the south shore of Waimea River. Scheffer wanted to use Kaumuali'i to extend Russian rule across Hawai'i, giving Russia a trade monopoly in the Islands. The fort originally covered thirty acres and had mounted cannons aimed seaward to protect Waimea. Kaumuali'i erected a home nearby, and Scheffer set up a factory, gardens, houses, and was served by Hawaiians. Scheffer's scheme to play off Hawaiian rulers against themselves, and to give the Russian czar the impression they were on his side, eventually came unraveled. He was forced to flee to Fort Alexander at Pu'u Poa point on a plateau overlooking Hanalei Bay adjacent to today's Princeville Hotel. On Kamehameha's orders Scheffer was sent packing, and left Hawai'i.

Missionary Settlements

Waimea was the last stop of the long voyage of the brig *Thaddeus* that brought

The star shape of the Russian Fort built by native Hawaiians is clearly visible from the air.

Outrigger canoes provided transportation and offshore fishing for native Hawaiians. The beachfront at Kalapaki served as a canoe landing in 1898.

the first American Protestant missionaries to Kaua'i. Earlier, visiting Tahitian Christians and explorer George Vancouver had brought word of the Christian God and his book, the Bible. Anxiously awaiting the end of the twelve-thousand-mile sail from Boston was George "Humehume" Kaumuali'i, the long-departed son of Kaumuali'i. Amazed and deeply touch at his son's return, Kaumuali'i welcomed the missionaries and immediately promised them lands at Waimea to start a mission station. The Reverend and Mrs. Samuel Whitney and their children set up a makeshift New England-style home at Hipo, on the coast adjacent to the Russian Fort. They were joined by the Ruggles family.

Kaumuali'i was among the first *ali'i* to become a Christian, and was the first to learn to read and write English with facility.

Missionary Peter Gulick toured the island in 1828, traveling from Waimea to Hanalei. Large crowds attended services, and it was decided that branch stations should be set up. The Wai'oli mission at Hanalei was founded in 1834 by the Reverend and Mrs. William P. Alexander. Their son Samuel became a co-founder of Alexander & Baldwin. There, a trade school opened with classes in carpentry, leather work, and Western farming methods. Another station was opened in 1835 at Koloa under the leadership of Gulick. The first Western doctors were stationed at Koloa, and in the 1850s Punahou principal Daniel Dole opened a boarding school for missionary children from all districts of the Island.

The first Roman Catholic mission to Kaua'i was established at Koloa in 1841 by Father Robert Walsh after an Edict of Toleration was issued by King Kamehameha III. Walsh built the stone church he named St. Raphael's in 1856. Another Catholic church was built at Hanalei.

Mormon missionaries arrived in December 1850 from the gold fields of California and found opposition from the Protestant missions and the Catholic church. J. W. H. Kauwahi helped complete the first translation of the Book of Mormon into Hawaiian at Waimea in 1854, and the church grew.

Above:
Humehume, George Kaumuali'i in an engraving made by Samuel F. B. Morse prior to the Kaua'i prince's departure from New England in 1819.

The grave of Samuel and Mercy Partridge Whitney is located in the churchyard of the Waimea United Church of Christ. Nearby mounds are said to contain the bones of Kaua'i ali'i gathered by Gov. Kaikioewa, who lived nearby in the 1830s.

Exile and the End of the Kingdom of Kaua'i

In 1821, Liholiho (Kamehameha II) and Kaumuali'i circled the Island, paying visits to villages. Kaumuali'i stepped aboard Liholiho's yacht *Ha'aheo O Hawai'i* when the party arrived back at Waimea. That evening the ship sailed to Honolulu, placing Kaumuali'i in exile until his death in 1824. His son Humehume, bitter over the circumstances of his life, realized that off-island *ali'i* would rule Kaua'i and take its lands. In August 1824 Kalanimoku, a wise advisor to Kamehameha, arrived to work out the land ownership. He told Debora Kapule and the Kaua'i *ali'i* that the lands would remain in their hands, but under the leadership of Ka'ahumanu and Liholiho, though some Kaua'i *ali'i* wanted to reallocate the *ahupua'a* of Kaua'i. In response, on August 8 Humehume and his forces attacked the garrison stationed at the Russian Fort. Humehume tried to rally support, and he called out from its walls, "Ho Waimea! Ho Makaweli!—come on—the Hawaiians are beaten—the Kauaians have the fort!" But no support came.

The Kaua'i force retreated to Wahiawa Valley. By August 20 reinforcements from O'ahu and Maui arrived to finalize the kingdom of Hawaii's hold on Kaua'i. Humehume and his followers prepared for battle at a spot near the overlook on the east side of Hanapepe Valley. The windward troops formed a right and left flank and advanced against men armed mostly with *o'o* digging sticks and wood spears. The defenders were slaughtered. Reports said between 50 and 130 were killed, including women and children. The histori-

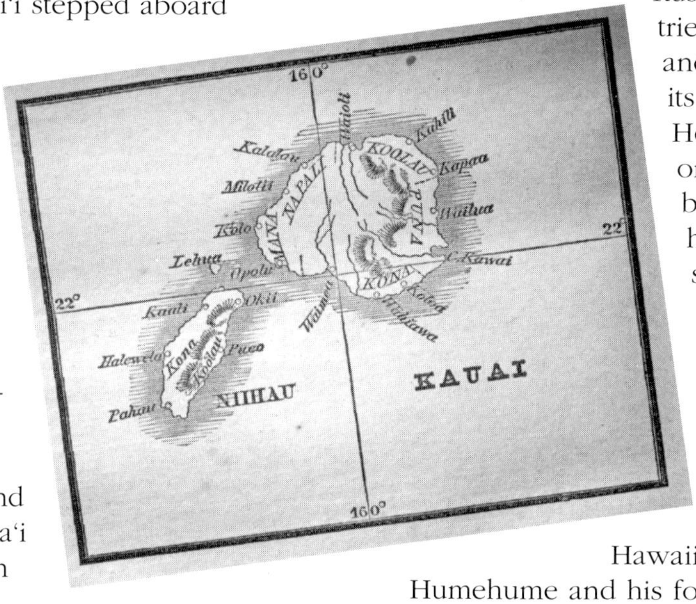

Above:
A missionary map published in New England in 1840 shows the location of principal ports and missionary stations.

Left:
Dr. James W. Smith served as Kauai's only western physician from the 1840s into the 1880s, often riding on horseback to care for patients across the island.

Whalers provided needed trade for Hanalei, and brought problems.

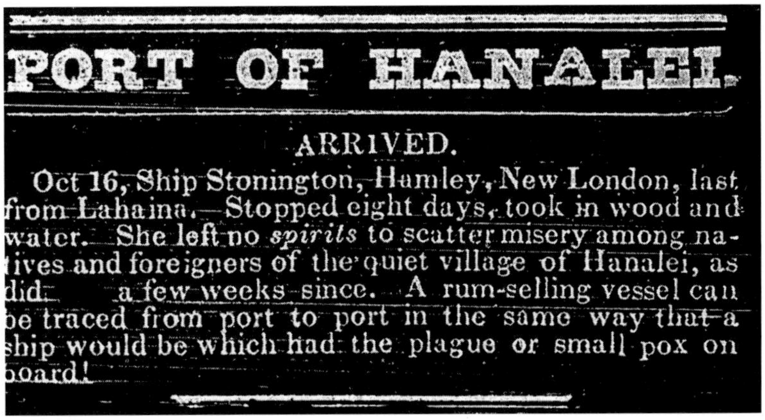

PORT OF HANALEI.

ARRIVED.

Oct 16, Ship Stonington, Hamley, New London, last from Lahaina.—Stopped eight days, took in wood and water. She left no *spirits* to scatter misery among natives and foreigners of the quiet village of Hanalei, as did   a few weeks since. A rum-selling vessel can be traced from port to port in the same way that a ship would be which had the plague or small pox on board!

an Kamakau called the battle "*Aipua'a*," or pig-eating, for the dead were left for the pigs to eat.

Humehume and his wife escaped, but were later captured and sent to Honolulu. In 1826 the exiled prince died of influenza.

By the end of the nineteenth century a number of large land tracts held by absentee landlord *ali'i* from other islands were sold for use as sugar plantations, ranches, and other developments. In 1824 Kaikioewa, a member of the court of the Kamehamehas, became governor of Kaua'i. He was the first of a series of governors to administer Kaua'i, until annexation of the Republic of Hawai'i by the United States in 1900 and the organization of county government in 1905. Lihu'e ranch and plantation owner William Hyde Rice was the last governor.

Whaling

Whaling began as sandalwood collecting ended. Whalers would hunt in the rich northern Pacific waters off Japan and North America in summer, and in the winter sail to the South Pacific. Between seasons they would re-provision on Kaua'i. Whalers watered and took on sweet potatoes, salted beef, oranges, barrels of butter, and other provisions at Koloa Landing, Waimea, Hanalei, Ni'ihau, and Wailua.

Koloa Landing was a favored port for it was out of the gusty trade winds and safe from large winter waves. Some years, forty to sixty ships called. George Charman and other traders at Koloa sent supplies to Honolulu for sale to whalers, including boxes of oranges and bags of coffee beans, melons, kegs of molasses, and firewood for firing try-pots at sea.

The infamous drunken whalers' riots in Lahaina and Honolulu were avoided on Kaua'i because there were no piers to tie up to, and crews were kept busy handling their ships as provisioning went on.

At Wailua, whalers anchored off the mouth of the Wailua River to load beef from the Brown estate located upriver. In 1854 the Honolulu firm of Hoffschlaeger and Stapenhorst acquired the lease to the estate and hired Scottish immigrant Duncan McBryde as manager. The German firm owned three whalers: the *Kauai*, the *Wailua*, and the *Victoria* which was probably named after Princess Victoria Kamamalu who owned large land sections south of Wailua.

The "singing sands" that give Barking Sands its name attracted visitors to the Mana Plain.

The inter-island sailing vessel John Ena *approaches Port Allen in this tile mural located in the board room of Alexander & Baldwin in Honolulu.*

Whaling peaked from the mid-1830s until 1861 when Confederate raiders began attacking Yankee whalers in the Pacific, forcing whalers back to New England. Demand for whale oil for lamps and lubrication dropped, too, as the use of petroleum, first discovered in Pennsylvania, increased. By the early 1870s sugar had overshadowed the whaling industry, which had suffered a death blow when the *Victoria* and other ships with ties to Hawai'i were stuck in ice in the Bering Strait and abandoned.

The Great Mahele

Ali'i and *konohiki* who controlled lands worked in cooperation with Kamehameha III in the division of lands known as the Great Mahele. Before, the rulers of Hawai'i held absolute title to all lands. Following the Mahele the *ali'i* retained control of royal lands, while the *maka'ainana* were given an opportunity to stake a claim for their family *kuleana*, or homestead. Traveling judges held court in outlying villages to take testimony for land claims that included statements from neighbors and *ali'i* regarding the claims. Public lands were also set aside. Some 937 claims were awarded by 1850, mostly to native Hawaiians, but also to missionary stations and missionaries, and immigrants like Frenchman and Kilauea rancher Jules Dudoit and Mexican *paniolo* Miguel Castro. The Great Mahele allowed the government for the first time to sell lands to the public, begin the surveying of Kaua'i, and start a real estate industry.

Government, Elections, and Taxation

In 1819 Kaumuali'i gladly

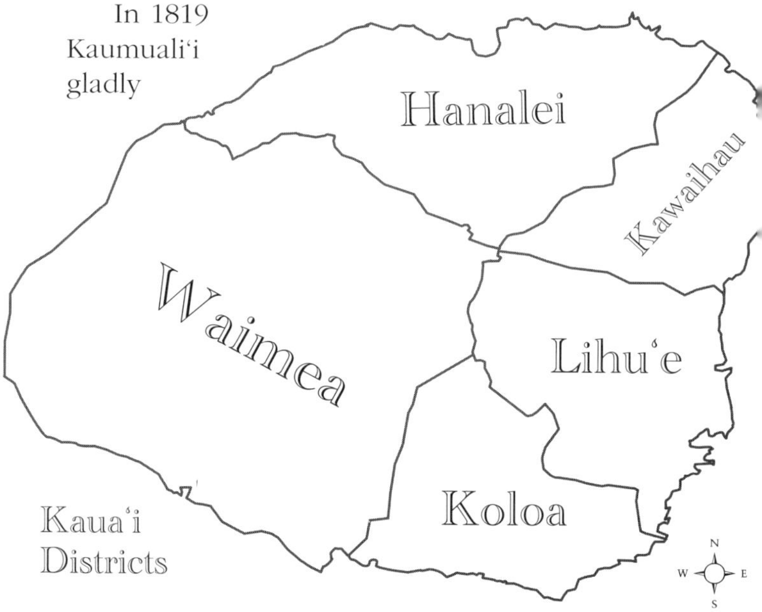

The modern districts of Kaua'i were drawn up beginning in the 1880s after King Kalakaua started up a sugar plantation mauka of Kapa'a, which he named Kawaihau.

The Birkmyre Estate overlooking the mouth of the Hanalei River is an example of a homestead built by a pioneering western family.

went along with the overthrow of the *kapu* system. This ended the ceremonial collection of agricultural produce, *kapa*, handicrafts, and sandalwood during the annual Makahiki season. He retained absolute control over the people and lands. The *konohiki*, or local overseers of each *ahupua'a*, were pressed into service for secular collections of taxes. Special taxes were also placed on the *maka'ainana* that forced them to collect sandalwood to pay for purchases of ships and cargoes of fancy goods purchased by Kaumuali'i.

Until Hawai'i became a territory in 1900 the island-wide gathering of taxes grew, with the imposition of tax on liquor, imports, and other use taxes, plus a one-dollar-per-year poll tax.

The Constitution of 1840 provided for a "representative body" of legislators, who were first chosen through petitions sent from Kaua'i to the king. The first popular election was held across the Hawaiian kingdom in 1851 to choose house members and road supervisors. A House of Nobles was appointed by the monarch until the Constitution of 1887 made both houses elective. Following the Overthrow of the Monarchy in 1893 the House of Nobles became the Senate.

Early Stores

Makeshift general stores and restaurants were run with a mixture of New England thrift, Chinese business sense, German order, and Hawaiian aloha. Some of the first stores outside plantations opened near the river mouth at Hanalei in the 1840s when Chinese merchants set up shop. The word *pidgin* is a corrupted form of the word *business*, and a word probably used by the Chinese merchants in dealing with ship captains, English planters, and the Hawaiians of Hanalei.

Set up later in the century, other shops with chinaware, bolts of cloth, hats, shoes, and tools for sale started as shacks thrown up on a plot of land on the edge of a port or plantation.

Barter abounded in the mid-1800s, and wasn't limited to shopkeepers. The ledger of Hanalei missionary Edward Johnson

The original Princeville Plantation begun in the early 1860s focused on fields along the Hanalei River. The sugar mill built by Robert Crichton Wyllie once stood downriver from the site of today's Hanalei Bridge.

shows that the Waioli Mission Station traded *palapala*—the written word in form of Bibles, tracts, and primers—for canoe paddles handcrafted in Kalalau Valley. The paddles were considered among the best in Hawai'i and sold in Honolulu to help support the mission.

Kaua'i money was invented by Koloa Plantation manager William Hooper, who set up a company store at Ladd and Company's Koloa Plantation, printing his own script to pay his workers. Supplies for the stores came from trading schooners, acquired for Spanish *reals*, and local produce, taro and rice, tobacco, dried fish, coffee, and bags of raw sugar.

In 1850 Lihue Store opened a few months after the granting of the royal deed for Lihue Plantation. The first post office in Lihu'e was housed in the store for fifty years, with general-delivery mail being handed out between sales. After 1876, the post office accepted deposits in postal saving accounts. That year the building was placed on rollers and moved to a site near where the county of Kauai's circular Mo'ikeha Building is located today.

Hardware and dry goods were the staples of the store, with a farmers market of sorts on Saturdays when fresh meat and produce were sold outside. Under missionary influence, Sabbaths were strictly observed, and workers were given a chance to shop on Saturday so they could prepare Sunday meals in advance. Though occasionally ice would arrive on ships from Boston, refrigeration was unknown. A quarter would buy a suckling pig, or a dozen mullet. Locally made products include *kapa*, rope made from sisal, and Panama-style hats woven from *lau hala*. Auctions with horses and sales items from the store featured were sometimes held on Saturday afternoons. In 1895 a larger building with a warehouse was constructed to replace the wooden building, with retail space rented to a tailor, a coffee shop, the first bank, and the first U.S. Post Office following annexation.

Ranching

The first cattle and horses were supplied to Kaumuali'i by ship captains. The cattle multiplied and roamed freely in the upland forests, but often came down and destroyed gardens and taro *lo'i*. *Papipi* (prickly pear cactus) was planted as a natural barbed wire to ward off the cattle. Missionaries and travelers often wrote in their journals of close calls with wild cattle. The best cattle and horses were taken off Kaua'i as plunder of war by *ali'i* of the Hawaiian kingdom in 1824 following the death of Kaumuali'i.

Initially, there was little commercial use for cattle, except for hides and tallow sometimes shipped off-island. Provisioning whaling ships created a market for salted beef.

Kaluaiko'olau, a paniolo at Waimea, and his wife Pi'ilani fled authorities who wanted to ship "Koolau the Leper" to the colony at Kalaupapa on Moloka'i.

Thomas Brown, the English coffee grower at Wailua, sold his ranch near Wailua Falls to Elard Hoffschlaeger and Florens Stapenhorst, two Honolulu-based Germans who sent three thousand barrels of beef to Honolulu each whaling season. They had seven hundred head of cattle at one time, and hired local men to roam from Kealia to Kilohana Crater to look for calves. During this era, meat went for two cents a pound in Lihu'e, and a good horse went for between ten and fifteen dollars.

European immigrants improved local stocks by bringing in purebred lines. The Royal Hawaiian Agricultural Society formed in 1850 and assisted in importing better breeds and livestock from the United States, Australia, and New Zealand. Valdemar Knudsen raised Longhorn, Durham, and Holstein cattle at his Waiawa ranch near Mana. William Hyde Rice raised cattle at Kipu, managing Lihue Plantation's ranch from 1867 to 1869. He started his own ranch in 1872, when the sugar plantations began to grow, to supply milk and meat for workers, forming the William Hyde Rice Dairy. G. N. Wilcox and Rice imported Herefords from Australia, and Ayrshires from New Zealand. Rice imported thoroughbred horses from Australia in the early 1880s and later shipped horses to Japan, including a Kaua'i stallion for the emperor. In 1884 Aubrey Robinson of the Robinson family at Makaweli imported Arabian horses

A glass bottle and waxed paper container from Waimea Dairy c. 1960. The dairy is remembered as the last locally owned dairy.

***Above*:**
An outing at the turn of the century brought together prominent Westside families including: Ruth Knudsen (Hanner), Margaret Russell Knudsen, Francis Gay, Lilly Hart Gay, Isabel Fayé, Augustus Knudsen, Ernest Gay (little boy), Margaret Lindsay Fayé, Anne Sinclair Knudsen, and a Japanese maid. H. P. Fayé is believed to be behind the camera.

***Below*:**
The Hawaiian Kingdom organized public schools on Kaua'i in the 1880s, often as extensions of schools begun by American Protestant missionaries decades earlier. In Kapa'a the first English-language school was opened for children of Portugese immigrants.

George Norton Wilcox. Missionary son, pioneering sugar grower, statesman, and philanthropist.

from Arabia and India. Brands were registered in 1848.

About the 1830s, *vaqueros*—Mexican, Indian, and Spanish cowhands—were hired to train Hawaiians in the handling of horses and cattle, roping, herding, and breeding cattle long before the cowboy tradition began in the western United States.

In 1831 Richard Charlton, the British counsel for Hawai'i, negotiated a twenty-year lease for lands *mauka* of today's Princeville Stables from Governor Kaikioewa. Charlton agreed to be responsible for shipping the livestock to Honolulu. About one hundred cattle grazed on the unfenced lands. Captain Jules Dudoit, the first French counsel to Hawai'i, took over the ranch lease in 1845 and moved his family over to Hanalei in 1848, setting up slaughter pens near where Hanalei Plantation Road is today. Dudoit's brig *John Dunlap* carried salt beef and cattle between Hanalei and Honolulu. In the 1860s a German named Krull ran an immaculate dairy farm on lands near Kealia Beach.

The domestication of animals led to the creation of dairy farms and ranches. Dairy cows pastured in drier, coastal areas like those of Koloa and Lihu'e, and the ranches were usually in cooler, upland areas, often areas unsuitable for sugar growing and near wild cattle. Cattle ranching increased as sugar plantations grew. The plantation found ranching a good use of marginal and fallow lands, and a good source of meat and dairy products for plantation workers and their families.

Cattlemen developed a close relationship with inter-island shippers, driving cattle to landings where they were swum out to steamers and schooners with holding pens fenced up on deck. The cattle were sent to feedlots in Honolulu.

Among the many successful businesses begun by Wilcox was the mining of sea bird guano on Laysan Island to the northwest of Kaua'i.

An 1874 portrait of Westside pioneers Valdemar and Anne Sinclair Knudsen and their family at Waiawa, the family's ranch once located north of Kekaha. The children are: Augustus, Ida, Maude, and Eric. Arthur was not yet born.

Reciprocity Treaty and Overthrow of the Monarchy

News of the fast-unfolding events that led to the overthrow of the monarchy in January 1893 failed to reach Kaua'i before the incident was over.

However, Kaua'i did have a role in the overthrow, with players on both the pro-monarchy and pro-republic sides. In October 1892 G. N. Wilcox was appointed minister of the interior by Queen Lili'uokalani, and William Hyde Rice as governor of Kaua'i. The "Wilcox Cabinet" is viewed as a last-ditch effort to head off the overthrow, but one that did not succeed. Days before the overthrow Wilcox was removed from office, and had no direct role in the overthrow.

In 1895 a counter-revolution on O'ahu led to the forming of a Citizens Guard on all major islands. Supplied with rifles and ammunition, the police-like guard set up bases at Makaweli and Lihu'e.

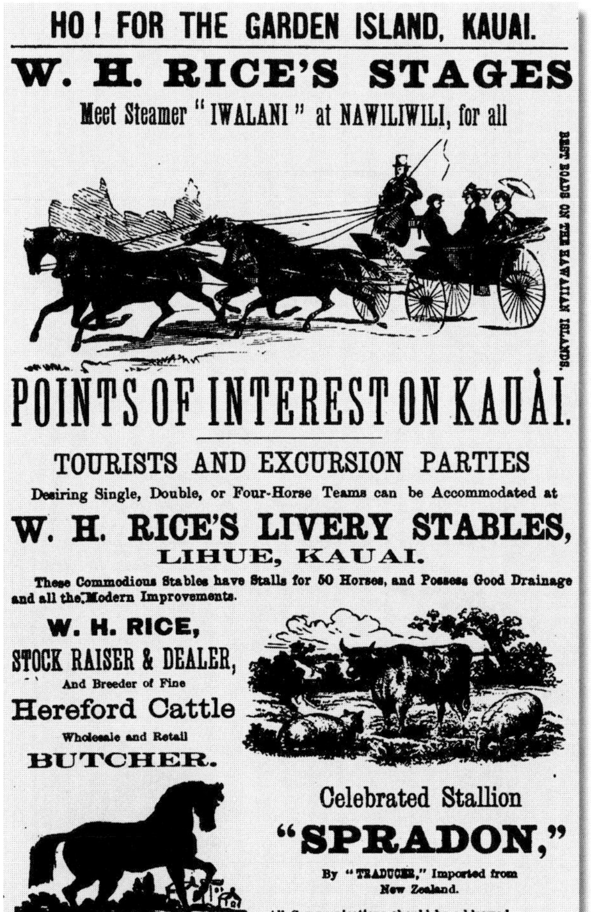

William Hyde Rice. Rancher, myth collector, and last governor of Kaua'i.

CHAPTER 3

The year 1900 was one of lasting changes for Kaua'i. The Island became part of the territory of Hawaii with the finalization of annexation to the United States. The disbanding of the Citizens Guard marked the end of that tumultuous era.

Electricity generated by hydro-turbines placed in streams brought the power for modern inventions like electric lights and radio which by 1930 would be commonly used. The first radio broadcast was heard in 1903. By 1908 G. N. Wilcox had the first automobile; there were 137 in use by 1912. Regular movie-going began at the Tip Top Theater, Kauai's first true motion picture theater, which opened in 1916. By World War II plantation theaters provided evening entertainment and a social gathering place in every plantation town.

With annexation the purse strings of federal territorial funding opened wide, adding to county funds for civic projects that brought Kaua'i a few milestones closer to the modern era. The funds spurred on the building of a macadam-paved highway from Ke'e Beach at Ha'ena about eighty miles to Mana. Nawiliwili Harbor opened in 1930 through the perseverance of G. N. Wilcox. Wilcox put up $200,000 in 1920 to keep the project alive, and to give Lihu'e the permanent mantle as the capital of the Island. The steel truss bridge over the Hanalei River was erected in 1912 along with other rural bridges, making unimpeded motor car travel across the Island. A U.S. Post Office opened in the Bank of Hawaii branch in Lihu'e, moving several times until the classical California mission-styled Lihue Post Office opened in 1930.

1900-1941

Congressional junkets arrived every few years, with interested congressmen enjoying the fruits of paradise while promising more modernizations, and even

(Continued on page 42)

The Kauai County government formed in 1905, and the first board of supervisors took the oath of office in 1906. (l. to r., back row) Arthur Rice, Treasurer 1905-1912; Olaf Omstead, Clerk 1905-1906; John Willard, Attorney 1905-1907; William Hyde Rice, Jr., Sheriff 1905-1942; J. Mahiai Kaneakua, Clerk 1906-1934; L. M. McKeague, Interpreter 1905-1906; S. K. Kaeo, Attorney 1907-1924 (l. to r. front row) Walter D. McBryde, Supervisor Koloa District 1905-1921; H. D. Wishard, Chairman Supervisors Lihue District 1905-1930; T. Brandt, Supervisor Waimea District 1905-1907; 1915-1924; D. Kanealie, Supervisor Hanalei District 1905-1907; R. Pukui (not pictured) Supervisor Kawaihau District 1905-1907.

Kaua'i: immigrant groups

Into the 1850s the ethnic makeup of Kaua'i was simply native Hawaiians, with a handful of resident American, British, and German pioneers, merchants, and missionaries, plus Chinese entrepreneurs running stores and milling cane at Waimea and Maha'ulepu.

In 1852 the Chinese began arriving as immigrant sugar workers from the Kwangtung province near Hong Kong, seeking a fortune in what they called Tan Heung Shan, or "The Sandalwood Mountains." They were brought to Koloa Plantation and Lihue Plantation as coolie laborers on five-year contracts.

The Chinese were allowed to migrate following the passage of a Masters and Servants Act by the Hawaiian kingdom that allowed the signing of labor contracts. Some stayed on as independent contractors on the plantations, or opened their own businesses, often hiring their own workers. Intermarriage to Hawaiian wives was common. Wong Aloiau and Hee Fat of Kapa'a became renown in the 1920s as wealthy "rice kings" with holdings of rice mills, stores, and warehouses full of rice grown in taro *lo'i* converted to rice paddies.

In 1878 Portuguese immigrants arrived from the Azores and Madeira islands. A civil war in Portugal depressed the economy of their home islands for decades, causing many families to seek a better life in New England, California, and Hawai'i. By 1886 there were 309 Portuguese working in the sugar fields of Kaua'i out of 1,988 immigrant plantation workers on the entire island. They were paid eighteen dollars per month. Many became key employees, often serving as lunas, and go-betweens to Oriental workers for their *haole* bosses. By 1888 the immigration was virtually halted due to the high expense of transportation from the faraway Atlantic islands.

Mass immigration of Chinese sugar cane workers first arrived in 1852.

Almost all Portuguese were Catholics. Their popular Holy Ghost Festivals featured fireworks, and colorful processions to High Mass. The Portuguese built dome-shaped *ferno* ovens to make *Pao Doce* bread. The Portuguese were avid homesteaders when interior lands became available in the decades after the passage of the Organic Act in 1900. Many settled in Kalaheo and Kapa'a, to enjoy a cooler climate similar to their homeland.

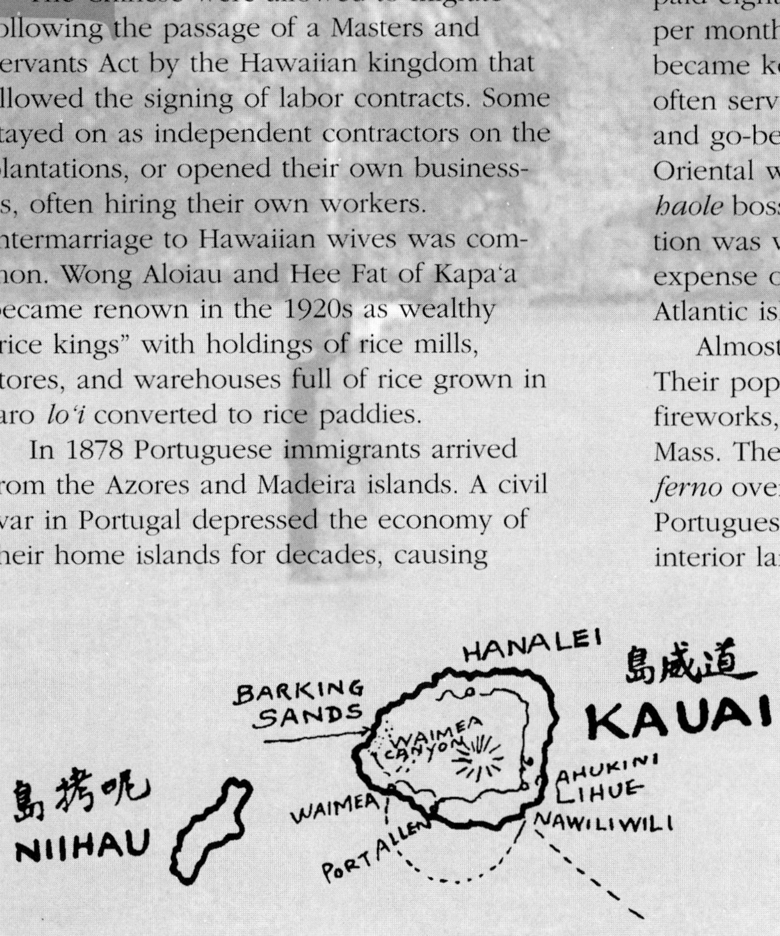

A Chinese map of Kaua'i drawn to commemorate the 50th anniversary of the immigrant groups arrival in large numbers in Hawai'i.

The sons and daughters of settlers from Madeira and Azores Islands found success on Kaua'i.

In 1885 the first significant numbers of Japanese immigrants began arriving, in an immigrant movement known as the *Kanyaku Imin.* Most of the men, women, and children came from southwestern Japan, from the rural farming prefectures of Hiroshima, Yamaguchi, Kumamoto, Fukuoka, and after 1900, the islands of Okinawa in the Ryuku Archipelago.

The *Issei* first generation of Japanese immigrants fled poverty in their villages as the Meiji Era began, ending the power of the samurai. They initially hoped to save up funds to purchase land back home. They received free passage by promising to serve three years in the sugar fields. They received firewood, lodging, and free medical care while working twenty-six twelve-hour days in the mill per month, or ten hours per day in the field.

Many Japanese immigrants chose to stay on, and some became independent contractors to the plantations. Generally, *Nisei* second generation moved on to become professionals, merchants, and government workers.

A high percentage of German immigrants to Hawai'i settled on Kaua'i. The German bark *Cedar* arrived in 1881, funded by Hackfeld and Company, with seventy-one men, nineteen women, and twenty-four children for Lihue Plantation. The immigrants were mostly from Hanover, a former independent kingdom west of the Elbe River, and driven to immigrate by poor economic problems and fear of military conscription into the conquering army of the Bismark. They worked at plantations with connections to Hackfeld and Company including plantations at Lihu'e, Koloa, and Kekaha. Today the German Hill neighborhood of Lihu'e recalls this era with its landmark Lutheran Church, the first one gathered in Hawai'i.

The German lunas ordered their workers around in a distinctive German-Hawaiian-English version of pidgin and made Christmas a merry season with decorated Christmas trees. Many later left for Honolulu and the mainland due to lack of land for sale.

Norwegian immigrants Valdemar Knudsen and members of the Fayé family

Japanese immigrants wait to begin their new life on Kaua'i plantations. Most Issei immigrants arrived between 1885 and the early 1900s.

Above:
As all immigrant groups did, the Filipino community formed social clubs like the YMCA Filipino club.
Left:
Port Allen was developed in the 1900s. The Westside port once rivaled Nawiliwili as Kauai's main port.

played a prominent role in sugar planting at Kekaha and Waimea.

Scots Robert Macfie, Jr., and Alexander Garvie came to work at plantations, and left behind the game of golf. Scots were recruited as plantation engineers and accountants. Most came from towns and rural districts in eastern Scotland between 1900 and 1909.

Puerto Ricans arrived in 1900 when that island became a possession of the United States. Famine and political unrest following the devastation of a hurricane drove them off-island. The Puerto Ricans took trains across the U.S. and brought wives and children with them. Their popular Kachi Kachi music rings out at local parties.

Koreans came as part of the Hawaii Sugar Planters Association's (HSPA) plan to balance Chinese and Japanese immigration to Hawai'i. A first wave of Korean immigration beginning in 1903 brought many Christians, recruited partly through the influence of American Protestant missionaries. The Koreans, like the Japanese, opened their own schools, like the Shin Heung Korean Language School in Koloa, community organizations, and churches. Later many moved to Honolulu, returned to Korea, or moved onto the U.S. mainland.

The first Okinawan immigrants arrived on Kaua'i in about 1906. By 1918 some 879 Okinawan men and 500 Okinawan women were on the Island, mostly working for sugar plantations. Okinawan communities gathered in Kalaheo, Kapa'a, and other towns near plantations. The Okinawans descended from a seafaring people, with roots in Korea and Taiwan, and immigrated due to hardships caused by overpopulation, a lack of jobs, and heavy taxes imposed by Japan. Okinawan organizations sent relief aid home following the devastation of their homeland during Allied-Japanese battles in World War II.

A reproduction of a traditional Filipino house built in the yard of the Wilcox family home at Hanalei entertained the children of Grove Farm plantation manager William Patterson Alexander.

Filipino immigrants arrived following the signing of the Gentleman's Agreement in 1907 that shut off Japan as a source of labor. The Philippines were a territory under the American flag, making immigration easier. Gradually the Japanese workers became displaced by Filipino workers. Most immigrated from the Ilocos region in northern Luzon. The Filipino immigrants were known as *sakadas* or indentured workers, and were actively recruited in their homeland by representatives of the HSPA. The Ilocos and eastern Visayan regions they came from were overpopulated. Early word came from Americans who had set up factories and agricultural enterprises in the Philippines that Filipinos made good workers. The plan was to apprentice young men to the plantations, train them in modern agricultural methods, and upon their return home they would help develop their own country. By 1925 recruitment was no longer necessary, as tales told from returning *Hawaiianos* helped keep immigrants coming.

From 1920 to 1929 a second wave of Filipinos arrived and by 1925 about 50 percent of workers on plantations were Filipinos. Union organizer Pablo Manlapit influenced the Filipino strikes of 1920 and 1924 that helped open the door to unionism on the plantations. Filipinos moved on from the plantations to pineapple plantations, docks, hospitals, hotels, and the armed forces, opened restaurants and small stores. In 1946 a postwar wave of Filipino immigrants filled jobs at plantations that stepped up sugar production after World War II, with 1,060 workers arriving at Port Allen. In 1965 a liberalization of immigration laws opened the door for modern waves of immigration, especially for those with relatives already on Kaua'i.

Smaller immigrant groups also added to the mix of life, including Spanish in 1907 aboard the *Heleopolis* from Malaga, South Spain, Pacific Islanders from the Gilberts and Ellice group, and even a short-lived try at farming by 150 urban Russian Molokans sent to Makee Plantation from California in the 1900s.

Above Right:
Immigrant workers ride to Gay & Robinson fields aboard cane train cars.

Left:
Chinese servants at Kekaha in the 1880s grew close to the sugar growing families they served.

The memory of Prince Jonah Kuhio Kalanianaole is honored each March by the Order of Kamehameha at his birthplace near Ho'ai Bay near Koloa. A plaque in his honor was dedicated on June 19, 1928 and drew a crowd of 10,000.

bringing over author Jack London in 1915 for a short visit.

In 1905 home-rule became a reality with the formation of the County of Kauai based on the American township system with a board of supervisors at the helm. The first supervisors, including pineapple planter Walter McBryde, met at Paulo Kanoa's old courthouse above Nawiliwili on a site to be occupied by Kauai High School once the County Building was completed in 1914.

Chinese and Japanese immigrants, and their sons and grandsons, prospered in the businesses they set up outside the plantation system. H. S. Kawakami in Waimea worked hard to develop his store. The family-run store was the beginning of the Big Save food store chain. W. Senda's photography studio and Kodak dealership brought him fame as Kauai's top photographer by the 1930s. The Silva stores in 'Ele'ele and Kalaheo served the Portuguese community and others. The Hanamaulu Cafe opened in 1923, owned by a Chinese family. German immigrant Christopher Hofgaard's general store at Waimea became the main gathering place in the town.

Norwegian immigrant Judge Christopher B. Hofgaard (1859-1931) ran a general store at Waimea that he founded in 1885 at the site of today's Big Save Market.

The shops of Kapa'a town were supported by workers from Makee Sugar Company at Kealia to the north of town.

Kama'aina families prospered, too, owning automobile dealerships and hotels, construction supply companies, and *The Garden Island* newspaper.

Annexation and its underlying federal Organic Act laws opened the door for petitions for homesteads to be made. Plantation workers wanting to go on their own found acreage in mauka Wailua and Kapa'a, at Lawa'i, and Kalaheo, awarded through a lottery system of land awards similar in spirit to the land rushes of Oklahoma, but on a much smaller Kaua'i scale.

The Chinese rice kings of Kapa'a and Waimea flourished, building rice mills, growing rice in the many abandoned Hawaiian taro *lo'i*. Their stores, which featured housing upstairs for proprietors, suffered a Chinatown fire in 1924 at Kapa'a that also devastated some schools.

The 1920s brought Kaua'i-style flappers at taxi dancing parlors, bootlegging arrests that embarrassed leading families, and flivvers for young men able to drive where they wanted on paved roads. Advances in communication drew the Island closer to Honolulu and the mainland through radio

(Continued on page 54)

Kaua‘i sugar plantations

The coming of sugarcane plantations brought a new major industry, providing work for residents of Kaua‘i, and a multitude of immigrants. The plantations and camps grew into today's modern communities, bringing improved medical care, electricity, and other improvements. Though each plantation shares a common heritage, each had a unique history and identity. Today sugar is no longer king on Kaua‘i. Only Amfac Sugar Kauai's Lihue Plantation and Kekaha Sugar Company, and Gay & Robinson, are still operating.

Koloa Plantation

Koloa Plantation is well known as the first commercial sugarcane plantation in the Hawaiian Islands. In 1835 the young Honolulu-based Yankee merchants of Ladd and Company chose 980 acres of land at Koloa for their plantation, later securing a lease from Kamehameha III for the site at three hundred dollars per year for fifty years. Plantation manager William Hooper struggled to build up the plantation, constructing a crude dam to run a mill that used koa wood grinders and water power from Maulili Stream. Hooper made up pasteboard "kala" script to pay the workers, redeemable at his company store on Saturdays. Oxen were used to haul the cane to mill.

The first sugar shipped in the second year of operation. Ladd and Company ran the plantation for twelve years, building up a model for plantation towns across Hawai‘i, but sold out to Dr. Robert W. Wood, the head of the American Hospital for Seamen in Honolulu. Wood built up the plantation and became in the 1850s the first successful sugar planter in Hawai‘i. In 1871 German immigrant Paul Isenberg bought a half interest in the plantation and became manager. Isenberg was a university-trained agriculturist. The plantation flourished following the passage of the Reciprocity Treaty in 1876, and despite having soil less rich than Makaweli or Lihu‘e, it moved ahead, providing employment for hundreds of workers until 1948 when it merged with Grove Farm. In 1974 Alexander & Baldwin's McBryde Sugar took over the mill and growing operation. The last harvest was made in September 1996 when McBryde Sugar shut down.

Paul Isenberg built Hanamaulu Mill to expedite the milling of sugar grown north of Lihu‘e. The mill closed in 1920.

Cutting sugar cane with machetes required large gangs of workers who often worked twelve hour days.

Lihue Plantation

Lihu'e town grew up around the mill and fields of Lihue Plantation, which was founded in 1849 by Henry Pierce, Charles Bishop, and William Lee. Earlier Governor Kaikioewa moved to Lihu'e, recognizing the potential of the coastal area, which he named Lihu'e after his homeland on the slopes of the Waianae Mountains on O'ahu. Lihu'e had a bay, ample rainfall, and many small streams.

Plantation manager William Harrison Rice is credited with beginning irrigation of sugarcane fields in Hawai'i, digging the first Lihue Plantation ditch in 1857 to bring water from Kilohana Crater down through Hanama'ulu Stream. Rice's ditch was two-and-a-half feet deep and ten miles long, and a wonder of its time. Irrigation allowed planters to not be dependent on rainfall.

Water power was added to its mill when the Nawiliwili Stream was dammed to form a mill pond. A few years later the plantation turned to steam power, and was the first to burn *bagasse* to heat water and create steam power.

Paul Isenberg joined the plantation in 1858 and went on to manage it and become an owner, incorporating the plantation in 1872, before returning to Germany in 1878. His interest in the plantation was overseen by his brother Carl Isenberg.

The smokestack of Kapaa mill stood as a landmark for the coastal east side of town.

Lihu'e Plantation brought in a forester from Germany to start the first reforestation project in Hawai'i. A railroad system was installed in 1891.

The original three thousand acres at Lihu'e were increased in 1866 by adding three hundred acres at Ahukini, and again in 1872 when a plantation at Hanama'ulu added seventeen thousand acres. In 1878, thirty thousand acres were leased at Wailua. A mill was erected at Hanama'ulu in 1877 to grind cane harvested north of the Hanama'ulu Gulch. Expansion continued with the purchase in 1910 of Makee Plantation at Kealia, and Princeville Plantation in 1916.

In 1922 American Factors acquired a controlling interest, and the mill was electrified using hydroelectric power. In 1933 Makee and Lihue Plantation merged operations, and the Makee milling equipment was moved to Lihu'e. Bulk shipping of sugar from Nawiliwili began in 1950. In 1995 Amfac Sugar Kauai East—as the plantation is officially known today—produced about seven tons of sugar per acre from forty-four hundred planted acres. The plantation now grows cane mostly in and around Lihu'e.

Grove Farm Plantation

Grove Farm Plantation is notable as the main plantation and funding source for the many successful investments made by George Norton Wilcox. In 1864 Wilcox leased four hundred acres of sugarcane lands from Judge Herman Widemann. The plantation's name came from a grove of trees that marked Widemann's home at Lihu'e. In 1870 Wilcox purchased the plantation and retained the name Grove Farm.

Wilcox's success at Grove Farm came through engineering innovations he applied following studies at Yale, especially in irrigation, and the use of fertilizer in his fields. His success as a sugar grower enabled him to play a leading role in community affairs and the economic development of Kaua'i.

The plantation grew from lands on the south side of Nawiliwili Stream to include 10,500 acres of Ha'iku lands, and additional acreage at Puhi. By 1935 Grove Farm was the largest plantation in the Hawaiian Islands that milled its cane at another plantation. In 1948, following Wilcox's death in 1933, Grove Farm bought Koloa Plantation, adding 3,041 acres to the plantation, bringing the company's land holdings up to 22,000 acres and adding a sugar mill. In 1974 Grove Farm went out of the sugar-growing business to focus on its land holdings, developments, and other businesses. McBryde Sugar Company leased Koloa Plantation from Grove Farm and grew and milled sugar there until 1996.

Hapai ko workers specialized in loading heavy bundles of cane into waiting narrow-gauge railroad cars.

Gay & Robinson

Gay & Robinson grows sugar on some of the most fertile soil in Hawai'i, and recorded a 14.67-ton-per-acre harvest in 1995. The company is looking ahead with plans to refine, package, and market its own brand of sugar.

Gay & Robinson's mill and sections of its plantation land, were once part of the former Hawaiian Sugar Company. Alexander

& Baldwin took a leading role in starting up the Hawaiian Sugar Company in 1889, after Sir William Renny Watson, a representative of a leading Scottish sugar equipment manufacturer, secured a fifty-year lease for Gay & Robinson land at Makaweli. By 1893 the plantation was made profitable by A&B's Henry Baldwin. His nephew Benjamin Baldwin managed the plantation for twenty-five years. The company built the

Many sugar plantations raised cattle to feed their workers. Here cattle are being rowed out to a waiting schooner or steamer as innovative Kaua'i boatmen use a cable and lighter to tow the herd offshore.

Olokele Ditch to increase the land available for cultivation. A&B's Kauai Railway Company spread a network of tracks throughout the plantation, including a trestle across Hanapepe Valley.

However, an agreement could not be reached on extending the land lease upon its expiration in 1941, and the Hawaiian Sugar Company was dissolved and its mill sold to the newly formed Olokele Sugar Company, whose agent was C. Brewer and Company.

Gay & Robinson is a family-run corporation with landholdings of 51,000 acres on Kaua'i. In 1994, Gay & Robinson purchased Olokele Sugar Company which is located in Kaumakani and now mills and grows its own sugar cane. The company continues to cultivate and preserve "plantation life and living" in three plantation camps found at Makawel, Pakala, and Kaumakani.

McBryde Sugar Company

The roots of McBryde Sugar Company go back to circuit court judge Duncan

Lihue Plantation along Nawiliwili Stream.

McBryde's widow Elizabeth McBryde, and her children. She formed a partnership with August Dreier that created Eleele Plantation, which had a mill at what is now Port Allen. In 1899, B. F. Dillingham merged Eleele Plantation, Wahiawa Ranch, and the Smith family's Koloa Agricultural Company Ltd., located on the Lawa'i side of Koloa, to create the McBryde Sugar Company. Alexander & Baldwin became the agent for McBryde Sugar in 1909.

Development costs were high due to the need to clear boulders from fields and the costs of securing a supply of water. Success came with the construction of a complex irrigation system that included thirty large and small reservoirs, most built between 1900 and 1907, bringing down

By the 1880s, steam plows moving at five miles per hour replaced oxen-pulled plows.

came to be called Numila. The old Eleele mill was sold off piecemeal and the new mill was noteable for its 84-inch cane crushing rollers, the largest in Hawai'i for fifty years. To power its mill McBryde built the Waniha hydroelectric plant in 1906. It was Hawaii's first and provided renewable energy for its pumps and other power needs.

Plantation workers were attracted to McBryde in the 1910s by the promise of homesteading land at Lawa'i and Kalaheo. By 1920 the plantation had fifteen hundred workers, company stores, and a hospital. In 1969 desilting ponds set up throughout the plantation led the way in pollution controls in the sugar industry. In 1974 McBryde grew from a thirty-thousand-tons-per-year plantation to the eighth largest in Hawai'i when Grove Farm Company went out of the sugar business. About 7,200 acres of cane land owned by Grove Farm and the Knudsen Estate were leased by McBryde, giving it over 13,000 acres under cultivation. In October 1974, the Numila mill with its landmark McBryde logo smokestack closed down and the company began milling at Grove Farm's Koloa Mill. In September 1996, after a long struggle with changing economic realities, McBryde stopped growing sugar and focused instead on the production of coffee, which its Kauai Coffee Company subsidiary had begun in 1987.

German immigrant sugar cane plantation owner Paul Isenberg played a leading role in developing Hackfeld & Company, the company that became American Factors in reaction to German involvement in World War I.

water from the streams of Wahiawa and Hanapepe. The construction of the Alexander Dam by McBryde beginning in 1930 is one of the most famous engineering projects ever attempted on Kaua'i. The mauka dam is still one of the highest hydraulic fill earth dams in the world.

McBryde's "New Mill" at Wahiawa opened in 1901, and the place eventually

Waimea Sugar Company

Waimea Sugar Company was an example of a small-scale, efficient sugar production, and grew cane in and around Waimea town. The roots of the plantation go back to early, unorganized plantings and milling by Chinese immigrants. Waimea missionary George Rowell exported five thousand pounds of sugar in 1854, collecting much of

The port of Waimea with inter-island vessels waiting offshore for cargoes of sugar cane from Kekaha Plantation, Waimea Sugar, and other Westside sugar growers.

the crop from small growers. A small mill was constructed in the 1860s. C. Conradt and W. Borehgervink began the planting of cane in Waimea Valley in 1880, and the Waimea Sugar Mill Company was formed in 1884. A mill was constructed and operated by H. Schmidt.

Because the salt content of underground water made it difficult to grow sugarcane, a ditch bringing water from the Waimea River was completed around 1900. The company began with less than 200 acres. About 1906 Hans Peter Fayé, manager of Kekaha Sugar Company, bought the plantation. A seashore marsh on the Mana Plain was drained, and the soil washed clean of salt, building the plantation up to 530 acres by 1935, with another 160 acres of cane grown by other planters being ground at the Waimea mill. A railroad was installed to speed up harvesting, and a tall, modern mill was constructed in 1916. During World War II sugar milling was moved to Kekaha Plantation and the arrangement continued after the war. By 1957 sugar yields topped fourteen tons of sugar per acre. The plantation's Waimea Dairy provided milk to Waimea and Kekaha. In 1969 the plantation shut down, following the death of Alan Fayé, Sr., whose career lasted from 1928–1968, and cane lands were leased to Kekaha Sugar Company.

Kekaha Sugar Company

Kekaha Sugar Company is now the west Kaua'i mill and plantation of Amfac Sugar Company, with lands running for about twelve

McBryde Sugar Company moved its mill from 'Ele'ele east to an area known as Numila.

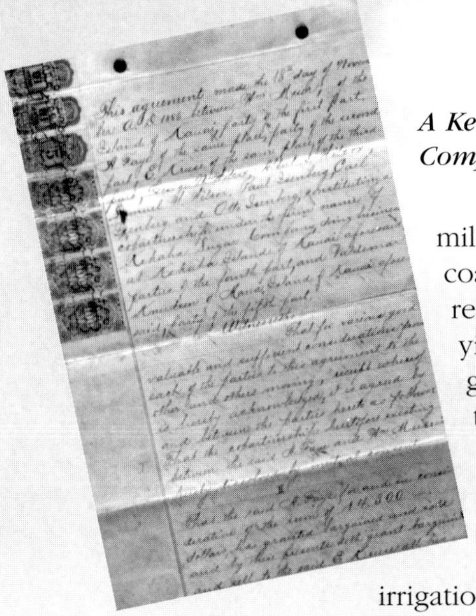

A Kekaha Sugar Company contract.

miles along the coast. A world-record per-acre yield has been grown at Kekaha thanks to a combination of abundant sunshine and careful timing of irrigation of fields by the plantation. Some mauka fields range as high as 1,850 feet up in the hills below Koke'e.

Kekaha Sugar was begun by Norwegian pioneer Valdemar Knudsen, who drained crown lands at Kekaha he had leased for thirty years. Knudsen planted about fifty acres of sugar in 1878 with Captain Hans L'Orange. In 1880 Paul, Otto, and Hans Isenberg joined with George, Albert, and Sam Wilcox in forming Kekaha Sugar Company. The company owned only a mill, which was operated by Hans Isenberg. Then, Valdemar Knudsen entered into a planting agreement with Kekaha Sugar Company. Anton Fayé and W. Meier bought out L'Orange in 1880. In 1884, Valdemar Knudsen made a new planting agreement with them doing business under the firm name of H. P. Fayé and Company.

By 1886 three operations were growing cane at Kekaha: German immigrants Meyer and Kruse cultivated lands around Kekaha; H. P. Fayé who cultivated the Mana lands; and the Kekaha Mill under the management of Hans Isenberg. An artesian well, based on an experiment on O'ahu, was dug about a decade later, and irrigation was begun. In 1887 Hackfeld and Company became the Honolulu agent for the plantation, an affiliation that continues today through Amfac Sugar Hawaii.

In 1898 the Kekaha Sugar Company was incorporated with H. P. Fayé as manager until his death in 1928. Bags of Kekaha sugar were sent by rail to Waimea Pier for export. By the early 1900s the plantation employed one thousand people and housed three hundred families.

In 1940 the plantation mechanized in response to a strike that took three hundred men out of the work force, and by 1947 rail operations were entirely replaced by cane-haul trucks.

The Fayé family provided managers for the company from 1897, when H. P. Fayé became the manager, followed by William Danford, and Lindsay A. Fayé, Sr., became the manager in 1935 and served until 1963. The last manager of Kekaha Sugar Plantation was Lindsay A. "Tony" Fayé, Jr., who served from 1980 to 1992, when the plantation began being operated at Amfac Sugar Kauai—West.

Princeville Plantation and Kilauea Plantation

In 1877 a tract of land bought in 1863 by Hanalei sugar pioneer Charles Titcomb became the Kilauea Plantation when

Three generations of Kekaha Plantation managers from one family: Hans Peter Fayé Sr. KSC manager 1898-1927, Lindsay Anton Fayé Sr. KSC manager 1934-1963, and Lindsay "Tony" Fayé Jr. KSC manager 1980-1992.

The Princeville Plantation mill once stood along the banks of the Hanalei River and featured state-of-the-art machinery imported from Scotland in the 1860s.

Captain John Ross and E. P. Adams purchased the property. Titcomb grew coffee and sugar in Hanalei Valley. His plantation was purchased by R. C. Wyllie, who expanded the lands and formed Princeville Plantation between 1861 and 1863. Sugar was milled at Wyllie's Glasgow-built mill along the Hanalei River. A number of planters unsuccessfully tried to make a profit there into the 1900s.

Kilauea had plenty of available water, but thin soil. The plantation installed the first railroad on Kaua'i in 1881, a narrow-gauge English system. "What was once vast unfenced pasture land is now a forest of sugar cane with a little town beside it," wrote a visitor to the opening of the railroad.

Kilauea Sugar Plantation Company was incorporated in 1899 and by 1935 owned 2,622 acres, and leased 2,226 acres.

Aerial cables transported bags of sugar milled at Kilauea Plantation to ships waiting in Kahili Bay.

Bags of sugar were shipped from Kilauea out of Kahili Bay, where a crane was built at a landing. The town of Kilauea grew up around a mill once located in the center of Kilauea. Camps dotted the plantation. Its fields ran mauka to Kalihiwai and south to Moloa'a. Distinctive stone buildings were a specialty of the plantation. The plantation flourished under manager David P. Larsen, a former head of field experiments for the HSPA, and his son, David W. Larsen. Maximum yields at Kilauea prior to its closing in 1973 were about six tons per acre.

Makee Plantation

In 1877 Civil War veteran Captain James Makee expanded his plantation holdings at Ulupalakua on Maui to Kaua'i. In a partner-

Anton Fayé faced harsh, arid conditions when he planted sugar at Mana.

ship with King Kalakaua he purchased the Ernest Krull dairy and pasture lands at Kapa'a, establishing the Makee Sugar Company. Makee and Kalakaua leased lands to friends of Kalakaua. Colonel Z. S. Spalding, Makee's son-in-law, acquired a majority interest in the plantation upon Makee's death in 1878. C. Brewer, and later Hackfeld and Company, served as agent for the plantation at Kapa'a. Irrigation came from a network of mauka streams. In 1885 Spalding dismantled the Kapa'a mill and moved it to his lands at Kealia. He kept milling work going at night by installing electric lights, making Makee the first mill in Hawai'i that could work after dark.

By 1889 1,030 workers were employed. Early homestead auctions at Kapa'a and Kealia went to workers. The Lihue Plantation Company, which earlier purchased all outstanding Makee stock, dismantled the mill in 1934 and moved it to Lihu'e. Remains of Makee Sugar include its barn-red, wood-sided office building and post office at Kealia.

In 1885 the Kapaa Mill and Makee Mill were combined at Kealia and an aerial tramway was installed to expedite moving sugar cane into the mill. The Makee sugar works dominated the landscape at Kealia.

Makee Sugar Company at Kealia used aerial cables to bring harvested sugar cane into the company-owned mill near the coast.

Kekaha Mill in the mid-1920s.

Above:
The H. S. Kawakami store at Waimea offered open-counter displays of merchandise, an innovation made in the 1920s. The pioneering Japanese-American merchant applied determination and merchandising know-how to create the successful Big Save grocery store chain.

Left:
A fire in 1924 devastated the heart of Kapa'a town, destroying buildings constructed with the wealth of local rice kings.

broadcasts and long distance telephone service.

However, Kaua'i retained its conservative ways throughout the era, reflected in some of the toughest liquor laws in the United States, and subsequently few problems with drunks.

The first signs of unionization of plantations began with sugar plantation and inter-island shipping strikes. The bloody Hanapepe Strike of 1924 resulted in the deaths of twenty-four, including strikers and policemen. In the late 1930s the ILWU organized dock workers at Port Allen as a key initial step to unionizing workers across Hawai'i.

In the 1930s the economy remained stable during the Great Depression thanks to a dependency on sugar, a staple food. The first airmail was delivered in 1934. A cornucopia of federal depression aid arrived to be

(Continued on page 63)

Left:
Business-sponsored boxing matches pitted local boxers against each other and proved popular after work entertainment in the pre-television days of the 1930s.

Below:
Local businesses brought leading national brands to stores in the 1920s.

The Tip Top Building, opened in 1915, provided Lihu'e with a modern anchor for its commercial center. The building was replaced by the "round building" of Lihue Shopping Center in the mid-1960s.

Kaua'i pineapple

James Dole perfected pineapple canning techniques and shipped his first harvest in 1903 from a plantation at Wahiawa on O'ahu. Within three years the industry spread to Kaua'i and quickly became the second largest export crop next to sugar.

By the early 1960s, four plantations were cultivating seventy-four hundred acres of land, or about 10 percent of pineapple acreage in Hawai'i, and were operating with an annual payroll in the millions. The eighteen-month crop was grown and canned by Hawaiian Canneries Company in Kapa'a, on a site now occupied by Pono Kai vacation condominiums, and by Kauai Pineapple Company in Lawa'i, in what's known today as the Old Cannery building. Pineapples mostly grown in Kapahi were brought to Hawaiian Fruit Packers. The cooperative processing plant was located in a corrugated warehouse building on Kawaihau Road. Grove Farm also grew pineapple, but sourced it out for canning.

The pineapple operations employed about two thousand summer workers including many high school and college students. By the mid-1970s all pineapple operations were shut down, mostly closed by competition from low-priced foreign imports.

Kauai Fruit and Land Company was the original name of "Kauai Pine" at Lawa'i, started up in 1906 by Walter McBryde and other Kaua'i businessmen as a subsidiary of McBryde Sugar Company. In 1910 Alexander & Baldwin became the agent for the pineapple company, which leased all its lands except for its cannery site, and accepted pineapples from homesteaders in Kalaheo and nearby towns. Following World War II the pineapple industry went into decline. In September 1964, all production ceased at Kauai Pine, and in 1965 all its equipment was sold.

Hawaiian Canneries opened in 1913 near the site of the original sugar mill in Kapa'a, and its fields spread out mauka to Kapahi and north to lands fronting the Kalalea Mountains below the landmark Hole in the Mountain. A signature logo for Hawaiian Canneries and its ties to major California food companies was widely promoted. Tins of the Kapa'a pineapples were shipped to Ahukini Pier for shipping by steamer to Honolulu and beyond. In 1956 some 1.5 million cases were packed in cans assembled next door by American Can. Its last crop was packed in 1962. Pineapples grown at Grove Farm and Princeville were sent to the canneries.

The Hawaiian Fruit Packers opened in 1932 as a cooperative venture of homesteaders in the Kawaihau area, and closed down in 1973. Today the sugar loaf variety of fresh pineapple growing is making a comeback at Kilauea. There, Mike Strong's Kahili Farms ships out a fine grade of the fruit to customers on the mainland.

Center: ***Pineapple pickers dressed in rough clothing and work gloves to harvest pineapples near Kapa'a in 1968.***

Kauai-grown pineapple was marketed across the United States under various labels including this Pono brand label.

Pineapple canneries provided seasonal work in the fields and cannery for college and high school students as well as extra income for families.

Hawaiian Canneries operated a pineapple cannery in Kapa'a. Today the Pono Kai vacation resort is located on the site of the cannery.

Aliomanu, the ahupua'a north of Anahola, was once planed in pineapple. The distinctive "Hole In The Mountain" visible in the Kalalea Mountains collapsed in 1983.

Kaua‘i railroads

Narrow-gauge railroads were technological breakthroughs for Kaua‘i sugar plantations, replacing hundreds of slow-moving oxen-pulled carts. The first plantation to have a railroad on Kaua‘i was Kilauea Plantation, which brought a complete railroad kit from England ’round the Horn in 1881.

Most of the plantation and common-carrier railroads of Kaua‘i started up in the 1880s and 1890s, with locomotives, cane-haul cars, track, and even coal coming mostly from faraway Germany and Great Britain. Portable rails moved along from field to field during harvest. Cane flumes deep in the mountains were used in areas too steep for locomotives to reach.

Sugar plantation owners and employees took pride in their trains. Many locomotives were given colorful names often after land sections and prominent citizens. A stipulation in the 1919 agreement with the federal government to develop a port at Nawiliwili called for linking together all the Island’s railroad systems so that any cargo from any point could reach the new harbor by rail.

By the 1950s most railroads closed, replaced by cane-haul trucks due to their maneuverability and lower operating costs.

Kauai Railway served McBryde Sugar Company, Hawaiian Sugar Company, and Koloa Plantation with a stable of several trains and more than eighteen miles of permanent track that spread out in an intricate network across the west and south coastlines. The rail line focused on bringing refined sugar to Eleele Landing, which became Port Allen in 1908. Koloa Plantation abandoned its old railroad line to Koloa Landing once it opened its new mill in 1912. Hawaiian Sugar Company’s railway dates back to 1891 when German immigrant August Drier imported German locomotives. In 1941 the company became the first in Hawai‘i to abandon railroads for trucks to move cane from the field to mill.

Ahukini Terminal and Railway ran from Anahola to Lihu‘e, and was organized in 1920 by American Factors to provide Makee Sugar with a connection to the dock at Ahukini Landing. The narrow-gauge, common-carrier railway was built between 1920 and 1921 heading north from Ahukini. An expensive wood trestle and concrete bridge were erected across Hanama‘ulu Valley. At Wailua River a 390-foot concrete bridge was built. The track continued north along the beach through Waipouli and Kapa‘a to Kealia. It ended at the old wharf on Anahola Bay. An extension to Nawiliwili was built in 1931 following the opening of

A wooden railroad trestle was erected across Hanama‘ulu Valley to transport sugar to Ahukini Landing for shipping to refineries and carrying cane to Lihu‘e.

The 24-inch gauge Fowler locomotive hauling cane at Kilauea Plantation is believed to be the first used on Kaua'i.

the port's modern docks. Today, traces of the old Ahukini line can still be seen across canals in Kapa'a and at the secondary bridge over the Wailua River.

Lihue Plantation was the last major plantation to replace oxen carts with locomotives, and the last to replace its railroad with cane-haul trucks. The rail line was extended to nearby Grove Farm Plantation in 1895, and in 1898 the small railroad of the Hanamaulu Plantation was added to Lihue Plantation. Grove Farm had several locomotives that were used to bring sugar in to be ground at Lihue Plantation. The cane was hauled to a wharf at Ahukini. A rail yard adjacent to the Lihue Plantation mill was used to service locomotives. Kekaha Sugar Company started rail operations in 1884, with rail running to Waimea Landing. Its railroad was mostly replaced by trucks by 1946. Waimea Sugar Company had a small railroad in the 1890s, and later built spurs mauka.

The whistle of a cane train was a common sound in plantation camps built near sugar mills.

To clear Westside fields for sugar planting required workers at McBryde Sugar Company and other plantations to move hundreds of boulders. Today the circular piles of basalt rocks can still be seen.

Upper Left:
Port activity at Nawiliwili once centered around a pier where lighters carried goods from schooners and steamers anchored offshore to waiting sugar growers, ranchers, wholesalers, and merchants.

Upper Right:
Rice Street circa 1920, looking mauka.

Center:
Lihu'e branch building (r.)

Clockwise From Top Left:

Garden Island Motors was Kauai's Ford dealer and located in the 1930s along the government road. The roots of the business go back to Nawiliwili Garage, founded in the 1910s along Nawiliwili Stream.

Wai'ale'ale and Kawakini peaks are the centerpiece of this territorial era County of Kauai logo.

KTOH went on the air May 8, 1940 as Kauai's first commercial radio station. Run as a subsidiary of The Garden Island *newspaper, the radio station had its own building and a broadcast tower along Ahukini Road where Jack Harter Helicopters is located today.*

Kauai Stores, first known as Lihue Plantation Store, closed in the early 1980s after serving as a general store in Lihue's town center since 1876. The store's signature sea captain logo is based on a wooden carved figure given to Mrs. William Harrison Rice by a sailor she nursed back to health.

used to improve roads, create getaway camps in Koke'e, improve ports, and set the stage for the arrival of masses of soldiers during World War II.

The 1930s brought an eight-hour work day and more jobs with workers laboring less hours per day at the plantations. Unemployment insurance, funded by pay withdrawals, provided temporary checks for laid-off workers beginning in 1938.

The plantation camps grew and improved, with neighborhoods of families becoming closer and closer as the decades passed. Companies outfitted baseball teams for inter-plantation play.

The era wasn't without deadly peril: automobile and cane train accidents regularly took lives; bouts of influenza and typhoid swept through towns; tuberculosis clinics were needed. The deaths of six workers—buried when the Alexander Dam gave way on the slopes of Mount Kahili—was the worst single disaster.

But the world still moved on. In 1937 the Rotary Club formed in Lihu'e. In 1940 young announcer Mike Ashman arrived to

Lihu'e town formed an L-shape in 1924. Commercial activity centered along Kuhio Highway and Rice Street, with the Kauai County Building erected in 1914, elegant homes, and sugar fields bordering the town center.

become a familiar voice on KTOH, Kauai's first commercial radio station. In letters home, Ashman told his parents back in San Francisco he couldn't see a world war actually coming to Kaua'i. By the next year he had packed his bags for a mainland interlude lengthened by the bombing of Pearl Harbor in December 1941.

Above Left:
Filipino immigrants brought professional music skills. This Filipino band broadcasts live over KTOH.

Above Right:
Editor and publisher Charlie Fern at the center of the action in The Garden Island Building on the makai side of Kuhio Highway in Lihu'e.

roads

Native Hawaiians traveled Kaua'i on foot along narrow trails that ringed and criss-crossed the Island. On Na Pali alone a complex series of trails hugged the shoreline and reached up to Koke'e. Most movement of sizable quantities of taro, dried fish, and other cargoes was done by paddling and sailing outrigger canoes. Smaller loads were carried in calabashes hung on poles and carried by a man.

Horses were introduced in the late 1700s, causing the trails to become wider. Riders forded streams and crossed rivers by canoe or flat-bottomed wooden scows into the early 1900s.

American Protestant missionary Peter Gulick walked from Waimea to Hanalei in 1827, and found two main trails passing Lihu'e: a foot path passing south of Kilohana Crater above Nawiliwili Bay and through Hanamaulu to the coast at Wailua, the other a wider horseback trail passing north of Kilohana Crater and reaching Wailua along the uplands.

By the mid-1800s horse-drawn buggies were in use. Cross-island trips in the new vehicles required skillful *paniolo* who would use their mounts to help pull the carriage up slick, muddy grades, and to serve as brakes on downhill runs, their lariats tied to rear axles. It took two days to travel from Lihu'e to Hanalei, with a stopover at Anahola at a government-owned guest house. As paths and horse trails became roads for the carriages, prisoners were sent out to do road work. Kaua'i was known as an island with excellent roads.

The coming of annexation in 1900 brought funding for a belt road of "water bound macadam roads" and bridges from Mana to Ha'ena. The first cars arrived via steamer in this era, numbering about three hundred by 1912, and creating a new industry for Island residents who opened automobile agencies, rented cars, repaired them, and offered taxi and tour services. By 1917, the belt road and several "laterals" heading mauka at Wailua and other areas, had been

A tree-lined dirt road at Makaweli used by horses and carriages.

Scows drawn by hand along cables stretched across rivers and streams allowed passengers and animals to stay dry at Lumaha'i.

paved, according to a report by County Engineer J. H. Moragne.

Some road sections followed coastal trails at Waimea; others were cut through cane fields between Lawa'i and Koloa. Moragne's main problem was keeping the early roads sealed, and removing rocks that rose through the pavement to puncture the narrow pure-rubber tires of early automobiles. He solved this problem for almost a decade by posting workers every two miles along the belt road to daily wet down the road and shovel red dirt atop it.

Automobile trips required ingenuity. A judge had to tie rope around his tires for traction to make it back to Lihu'e from Hanalei, and visitors used a pick and shovel to get their car up to Koke'e. An open car took two and a half hours to make the climb.

By 1920 Nawiliwili Garage and other early dealerships sold Ford Model-Ts, Oldsmobiles, and other brands. Two small boats lashed together into a catamaran were used to land cars. In 1924, Kawaihau Road was constructed for homesteaders. In 1927 the road was paved to Ke'e Beach, allowing tourists to see the famous Ha'ena caves, and in 1929 a paved

Early touring cars were grand objects and oddities running along the dirt roads of Kaua'i in the 1910s.

Building of a "belt road" from Mana to Ha'ena required surveying, grading, and a paving method known as macadamizing. Here a road is prepared outside Hanalei.

road reached the plateau above Wailua Falls.

A rising number of accidents caused highway officials to put up the first stop sign on Kaua'i in 1937, near the intersection below Tip Top Store. The first traffic light was put in by Kekaha Sugar, for cane-hauling trucks crossing the road to Koke'e. Federal Depression–era funding put men to work painting solid and dotted lines for passing and no-passing zones. Street naming was an ongoing event. Main drags in Lihu'e once had names like Elite Street before they were named after William Hyde Rice, Judge Jacob Hardy, and for where they led. In 1938 the county board of supervisors approved naming the western side of the belt road after Kaumuali'i, the last king of Kaua'i, almost allowing the highway to be named Roosevelt or Kamehameha.

World War II manpower and massive imports of excess war materials brought about improvements to roads and bridges. In 1954 state senator Miyake of Waimea sponsored construction of a mountain-top road from Koke'e to Ha'ena that ran along Na Pali cliffs. After a three-mile gravel jeep road was constructed, funds dried up. In 1963 the multi-million-dollar Kalihiwai Bridge became the longest in the state, running eight hundred feet over the Kalihiwai River and replacing a makai truss bridge washed away by the 1957 tsunami.

Center:
Fording a stream on horseback was a common occurrence for cross-island riders in the 1890s.
Right:
Paving over ancient rails allowed drivers to visit homes at Ha'ena and other remote valleys once accessible only by canoe, horseback, or by foot.

A crossroad allowed travel to Lihue Mill from Lihue town and Koloa.

The arrival of 1970s and 1980s development drove the creation of new roads, four-lane widenings, and installation of improved signs and other safety features. The Ahukini By-Pass opened in the late 1980s, and other bypass roads helped improve the traffic flow of fifty-thousand-plus automobiles.

Above: *The automobile made isolated coastlines accessible to the public. A swimming platform at the mouth of the Kilauea River drew a fleet of Model-T's in the summer of 1924.*

Left: *The horse and buggy were introduced in the 1850s, allowing more comfortable transportation between Waimea and nearby towns.*

By the 1930s a paved road comfortably carried cars and trucks through Waimea town.

Kaua'i shipping

Since the first settlers sailed canoes to Kaua'i over a millennium ago, the sea has been a vital link to the outside world.

Today container barges, towed by tugs across the Kaua'i Channel by Matson Navigation Company, Sea-Land Service, and Young Brothers, economically and safely keep this vital link open, bringing in automobiles, groceries, lumber, and thousands of other vital cargoes.

The arrival of Western explorers in the late nineteenth century made sailing between islands more common. By 1820 inter-island schooners were commonly seen anchored off Waimea, captained by adventurous New Englanders loading cargoes of sandalwood and provisions.

Whalers from New England and Great Britain made Koloa a main port during the whaling era, which dovetailed with the ending of the sandalwood era in the 1830s. Crews faced stiff trade winds, high winter seas, and faced going aground on reefs and off unprotected landings. Low-slung whale boats known as lighters were used for loading and unloading cargo. It was hard work, slow and sometimes dangerous. There were no docks to tie up to, just landings in shallow water. Sometimes cargoes were landed through the surf.

"hand along the smooth leaf brings back memories of old Hawaii."

INTER-ISLAND STEAM NAVIGATION CO.

STEAMER PLANTER

STEAMER IWALANI

STEAMER C. R. BISHOP

STEAMER JAMES MAKEE

FOR SALE

COAL, IRON AND STE

A broadsheet posted in Honolulu offering passage to Kaua'i aboard a steamer.

Passenger travel could be perilous. In March 1858 the schooner *Excel*, commanded by Captain Antonio, sailed for Honolulu in stormy weather. Ten days later the ship was discovered north of Maui almost out of water and food.

However, inter-island schooners were used to see family on other islands, with passengers paying between fifty cents and one dollar for deck passage limited only by how many bodies could be crammed on board.

A schooner off the pier at Ahukini, the first port on Kaua'i where passengers could land directly on shore.

Steamers and schooners transporting passengers and cargo to and from Nawiliwili stood well off the coast until 1930.

The trip between Honolulu and Koloa took up to forty hours, and malihini passengers were advised: "Foreign passengers on both of these vessels will require to provide themselves with provisions unless they choose to put up with fare a la Hawaii."

In 1853 the steamer *Akamai* came into service and shipped meat, butter, and oranges to markets in Honolulu, usually taking sixteen hours to cross the channel. The coal-driven steamer could sail in any direction.

The names of the Kaua'i-run vessels were popularly known: the schooner *John Young*, the eighty-seven-ton *Annie Laurie*, the brigantine *Kilauea*.

Following the Civil War, new laws required a vessel inspection, the carrying of extra barrels of food and water, and licensing of captains who could find longitude and latitude. Women under the age of twenty-five from the outer islands were required to acquire a passport "signed by a person of authority" as a measure against importing innocent Kaua'i girls to Honolulu to serve as prostitutes. Setting channel and anchorage buoys, and the beginning of a kingdom-wide lighthouse service, were other improvements.

Loading cattle at outer-island landings such as Ki'i on Ni'ihau was expensive for steamship companies, with paniolo swimming cattle out to waiting ships, and deck hands using a huki (lift) to haul the animals on board. The operation took hours and required extra crewmen. Cattle were placed on deck in specially designed pens.

The Reciprocity Treaty of 1876 brought great growth in the shipping of sugar and rice to Honolulu for trans-shipment to California and New York for refining.

The kingdom funded 592-ton steamer *Likelike*, which could carry larger cargoes.

In 1883 Inter-Island Steam Navigation Company was formed with G. N. Wilcox, a main investor, bringing regularly scheduled port docking dates for passengers and shippers.

The ships of Swedish-born Captain William Matson began taking California-bound sugar from Kaua'i in September 1908 for factors, and returned with cargoes for businesses. Castle and Cooke, Alexander & Baldwin, and C. Brewer and Company bought stock. Hackfeld and Company, the German sugar factor with ties to Lihue Plantation, ran the

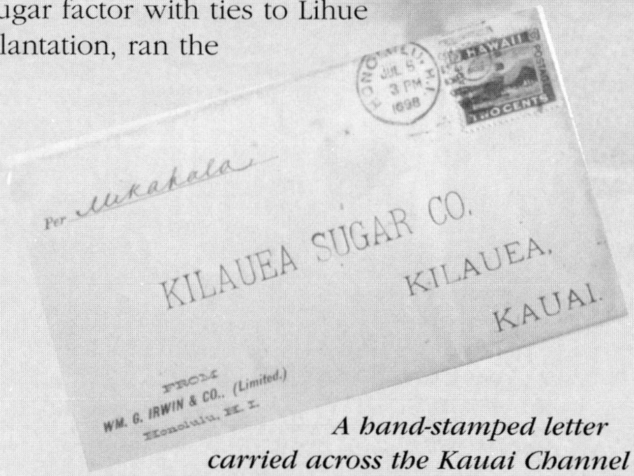

A hand-stamped letter carried across the Kauai Channel aboard the steamer Mikahala.

Kaua'i photographer Edward Deverill captured on film the shipwreck of the SS Ivanhoe. *The freight vessel wrecked in 1915 and its fate was shared by many vessels sailing in Hawaiian waters. The once handsome iron-hulled ship was first used to send emigrants to Australia. It was a tramp coal carrier when it went aground.*

imperial German flag aboard their ships *G. N. Wilcox* and *Robert W. Wood.*

Inter-island ports and landings included Ki'i and Lehua on Ni'iahu plus Nawiliwili, Koloa, Eleele, Waimea, Ahukini, Hanama'ulu, Kapa'a, Kealia, Anahola, Moloa'a, Kahili below Kilauea, Kalihiwai, Hanalei, and Wainiha.

The arrival of the *Mauna Kea* brought the first oil-fueled ship to Kaua'i service, and offered a grand passage with promenade decks and bridal suites. Steerage passengers were fed in a unique "poi room."

In 1913 the Territorial Public Utilities Commission brought further regulation to inter-island shipping, and as tourism grew Matson added the *Lurline* and *Matsonia* to the West Coast service.

G. N. Wilcox became president of Inter-Island Shipping in 1924 at the age of eighty-five. In 1925 the 210-foot diesel-electric ship *Hawaiian Standard,* an oil tanker, began delivering Standard Oil products to bulk storage plants at Ahukini and Port Allen. In 1928 the 400-foot-long *Waialeale* went into operation, named by G. N. Wilcox to help promote Kauai. Staterooms had private toilets, showers, mahogany furnishings, and a radio. The Inter-Island Steamship Company also handled coal sales, chandlery, ship repair, travel agency, office buildings, hotels, and air transport.

The federal government requisitioned many inter-island ships following the outbreak of World War II. The ships alternated between carrying troops and supplies to Palmyra, Fanning, and other far-flung Pacific ports, and providing irregular service to Kaua'i. Flying to Honolulu became popular, and air fares became competitive. By 1947 only 3 percent of passengers crossing the Kaua'i Channel went by boat.

Center:
A sailor rows passengers out to a waiting steamer at Nawiliwili in a drawing made by a visiting artist in the 1890s.

Passengers at Nawiliwili dress up for a channel crossing.

Steamers replaced schooners in the inter-island trade, making voyages between O'ahu ports and Kaua'i ports safer and faster.

A railroad running from Polihale to Waimea Landing carried sugar, rice, and other produce for shipment off island from Waimea Pier.

Eleele Landing grew into Port Allen. Native Hawaiians once used the site as a canoe landing.

Nawiliwili Harbor was officially opened on July 22, 1930 with the docking of the inter-island steamer Hualalai, offering a deep water protected port for large ships.

Young Brothers expanded its inter-island freight service to Kaua'i in 1947, offering biweekly service to Nawiliwili and Port Allen. The company considered itself an extension of the common-carrier railroads of the Islands, and was founded by adventurous young brothers William, Edgar, Herbert, and John Young of California. Edgar Young later served as a Makaweli Plantation doctor.

Common-carrier tug and barge service cost less than a steamer of equal capacity, used smaller crews, and cargo could be preloaded. By 1951 Young Brothers Limited tugs were making three trips per week to Kaua'i carrying twelve hundred to fifteen hundred tons of cargo.

In the late 1950s Matson pioneered the use of large containers when it shipped cargo in them to Hawai'i. Container shipments cut handling costs drastically, reduced cargo damage and pilferage, and shortened the turnaround time for vessels. By 1963 Matson was delivering containers directly to Nawiliwili aboard the three-hundred-foot-long barge *Islander*. The barge returned to Honolulu with exports of pineapple, household goods, and other mainland-bound cargo.

air transportation

About a million airline passengers each year fly in and out of Lihue Airport. Visitors from the mainland and countries across the globe, plus local residents, take for granted the excellent service provided by Hawai'i-based and national airlines. Support for expansion of the main runway of the modern jetport and arrival of larger aircraft points to an increase in those numbers.

The roots of the aviation industry date back to January 1914 when aviator Tom Gunn made two short flights from a field in Koloa near Po'ipu Beach, drawing fifteen hundred spectators who watched a five-minute and a three-minute flight. In May 1920 *Garden Island* publisher and World War I pilot Charlie Fern flew in as a barnstormer, hawking rides following ome of the first cross-channel flights to Kaua'i.

In 1925, Lieutenant Commander Rodgers and his PN9-1 seaplane crew were rescued near Kaua'i and towed into Nawiliwili Harbor to complete the first mainland-Hawai'i flight, creating a sensation, and news stories with a Kaua'i dateline appearing for the first time across the globe. That day an elderly native Hawaiian man repeated the Hawaiian proverb: "All good things come to Kauai at last."

An Inter-Island Air Sikorsky S-43 at the Burns Field terminal. The planes went into service beginning in 1935 after being shipped to Honolulu by freighter from the West Coast.

Charles Kingsford-Smith landed the *Southern Cross* for a fueling stop at Kekaha Sugar's Barking Sands airstrip in May 1928 on the first flight from the West Coast to Australia.

The first scheduled passenger flights began on November 12, 1929, with the arrival of Inter-Island

Hollywood actress Carmen Miranda arrives aboard a Trans-Pacific Airlines flight in the late 1940s. The upstart airline later became Aloha Airlines and first flew charter flights to an airfield at Barking Sands.

Airways, Limited's twice-weekly flights to Burns Field in Hanapepe, flying a Sikorsky amphibious aircraft. The plane held two pilots and eight passengers, and needed a two-thousand-foot runway to land. The pontoons mounted on the wings were for emergency landings while crossing the wide Kauai Channel. Inter-Island investor G. N. Wilcox was onboard, and enjoyed a side trip over Waimea Canyon before landing.

The first airfield at Lihu'e was opened in 1934 just north of Hanama'ulu Bay, but because of its short runway most flights went to Burns Field and Barking Sands until a modern airport was opened between Ahukini and Nawiliwili in 1950. Airmail service began to Kaua'i on October 8, 1934. Trans-Pacific air passengers making trips to Kaua'i picked up when Pan American began its clipper service to Hawai'i in 1936.

On October 1, 1941, Inter-Island Airways, Limited became Hawaiian Airlines, Limited. Months earlier new Douglas DC-3 aircraft arrived to replace the aging seaplanes. The coming of World War II placed all inter-island aircraft under military control. The War Shipping Administration decided that Hawaiian Air would be the only public means of transportation between the Islands, and only flew during daylight hours. The DC-3s were used for passengers and the Sikorskys carried freight. A round-trip ticket to Honolulu cost $24.30.

When inter-island ships were rerouted to war service, the public became accustomed to flying between islands, causing Inter-Island Steamship Company to discontinue passenger service in 1948.

In 1947 the site of the 210-acre Lihue Airport was chosen with a 4,600-foot runway intersecting a 3,600-foot runway at "the junction of the main Ahukini Road and the Lihue garbage dump road" on Lihue Plantation land. In September 1949 the Civil Aeronautics Administration landed the first airplane at the new airport. The airport opened for regularly scheduled flights on January 8, 1950.

Entrepreneurs saw an opportunity to challenge Hawaiian Air after the war, and in late 1946 Dr. Hung Wo Ching and Rudy Tongg launched Trans-Pacific Airlines. The Hawaii-owned company began to charter inter-island air service with unpressurized DC-3s still in military configuration. Passengers sat in metal bucket seats running like benches along the fuselage facing inward. Some three hundred people showed up to greet the kickoff flight to Barking Sands field. Following years of legal and marketing battles, in 1951 TPA Aloha Airlines gained permission for scheduled flights and airmail routes. The company grew into Aloha Airlines, flying DC-3s for fifteen years.

In the postwar years Hawai'i was still considered an international destination, and flight attendants served pineapple juice in-flight, sang, and played ukulele.

The competition between Aloha and Hawaiian Air forced the companies to bring

Inter-Island Airlines launched passenger air service to Kaua'i on November 11, 1929. This advertisement appeared earlier that year in **The Garden Island*, on August 13.***

in the latest passenger planes. The last DC-3 flew in 1966, the year inter-island jet travel began. Hawaiian Air introduced DC-9 jets that could fly at thirty-five thousand feet and carry over one hundred passengers as tourism flights boomed. Another innovation was the DC-933RC, a plane that could be switched from cargo to passenger configuration in thirty minutes. Hawaiian and Aloha air cargo also provided competition for Matson and Young Brothers.

By the 1970s inter-island jet travel was more common than bus rides on Kaua'i, and the industry grew as resort hotels were developed.

In 1986 a new terminal with jetways that connected the terminal directly to aircraft opened, named for the late Kaua'i legislator Richard Kawakami. In 1998 United Airlines resumed direct flights from the West Coast to Lihue Airport after a six-year shut down following Hurricane 'Iniki. The charter helicopter companies flying out of the state of Hawaii's heliport adjacent to the airport freight terminals, plus helicopter operators flying out of Burns Field at Hanapepe and Princeville Airport, constitute the largest fleet in Hawai'i.

A post-war Hawaiian Airlines DC-3 flight at Lihue Airport.

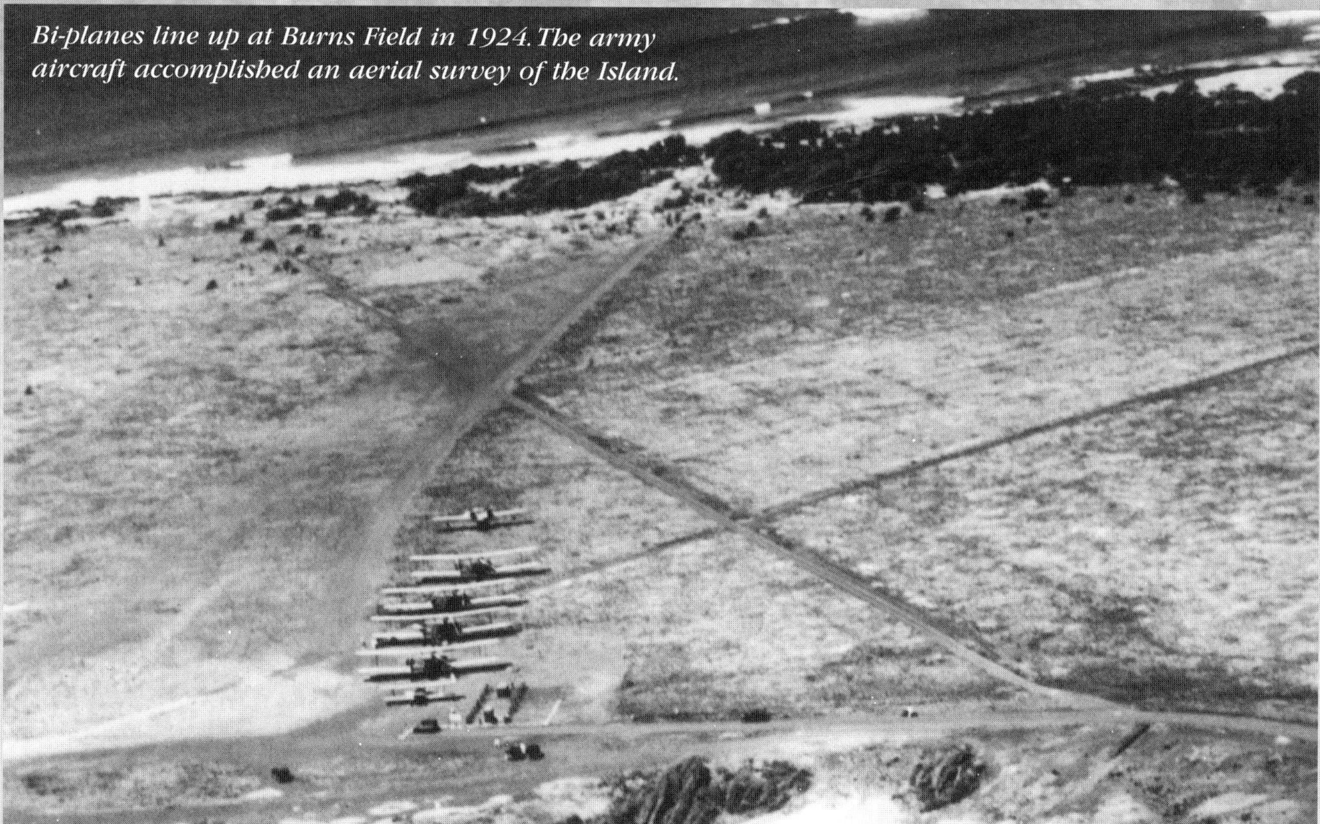
Bi-planes line up at Burns Field in 1924. The army aircraft accomplished an aerial survey of the Island.

medical care

The medical care industry is one of Kauai's largest employers today. Its roots go back to ancient Kaua'i. Then the Island was isolated from many diseases, and native practitioners known as *kahuna lapa'au* used native medicinal plants, sea salts, prayers, and other remedies to treat native Hawaiians.

The coming of Captain James Cook in 1778 brought the first of a wave of epidemics that would decimate the native Hawaiian population over the next 125 years. Against Cook's orders, his sailors spread venereal diseases on Ni'ihau. Later arrivals of European and American whalers and merchant ships brought epidemics of cholera, smallpox, whooping cough, and other ravaging diseases, plus introduced the common cold, pneumonia, and mumps. In the 1850s workers imported from China for plantations introduced leprosy.

American Protestant missionaries were the first Western settlers to battle these diseases. An early medical book in the Hawaiian language, *Anatomia* by missionary Dr. Gerrit Judd, was made available in the late 1830s. In 1843 missionary Dr. James W. Smith arrived at the Koloa missionary station, and was the only Western doctor for forty years. Smith fought off a smallpox epi-

The first twins born at G. N. Wilcox Memorial Hospital.

demic in 1853–1854 by widespread vaccination, and traveled by horseback on emergency calls, and to see patients in towns and villages.

By the 1850s plantations began offering health care to their growing number of plantation workers, both to help the workers and keep plantations running. In 1880 G. N. Wilcox founded Kauai Telephone Company to assist Dr. Smith in reaching patients in Lihu‘e and other towns without traveling by horseback.

Plantation-sponsored medical care brought healthier lives to more people, and the first plantation hospital in Hawai‘i was opened at Hawaiian Sugar Company in Makaweli in 1886. Several years later hospitals opened at Makee Plantation in Kealia, at Kilauea Plantation, and at Waimea Plantation. Plantation clinics became social centers, as well as dispensers for medicine, first aid, and general care.

The Mokihana Club hired a district nurse, and worked on sanitation programs for plantation camps by 1920. Nursing training required a move to Honolulu for classes at Queen's Hospital nursing school, which opened in 1916. The spread of tuberculosis in the 1900s resulted in the opening of a sanitarium at Kapa‘a in 1917, funded by the county of Kauai and the Wilcox family in memory of Samuel Mahelona Wilcox, a young man who died from the disease.

Elsie and Mabel Wilcox, their uncle Grove Farm Company owner G. N. Wilcox, Gaylord P. Wilcox, as well as Dora Rice Isenberg, worked diligently and throughout their lives to improve the medical care of Kauai's people. In 1938 the G. N. Wilcox Memorial Hospital opened in Lihu‘e, built on land donated by Lihue Plantation. The hospital brought a continually growing number of modern medical services to the Island including a registered pharmacy, a medical lab, various surgeries, and a centralized location for private doctors and plantation doctors to provide treatment beyond that available at plantation clinics. In 1957 the state-funded Kauai Veterans Memorial Hospital opened at Waimea. In the 1960s nursing degree programs were offered at Kauai Community College, allowing local residents to study the nursing field without travel to Honolulu or the mainland. The hospitals have driven the introduction of many modern technologies to Kaua‘i, including advanced computer systems, CAT-Scans, and MRI technology. With the coming of the new century the hospitals continue to play leading roles in the community as centers of employment and care for all.

G. N. Wilcox Memorial Hospital was dedicated on November 1, 1938 on a plateau in Lihu'e overlooking Kapaia Valley. Land for the hospital was purchased from Lihue Plantation.

CHAPTER 4

Kaua'i was somewhat prepared for World War II, yet still shocked when it came on December 7, 1941. For months the Office of Civilian Defense, led by *Garden Island* publisher Charlie Fern, other community leaders, and plantation officials, had begun wartime preparations with plans for a local militia and evacuation of the civilian population into the interior if an invasion occurred.

It wasn't until after the war that common knowledge of a "scorched earth" policy was reported. According to World War II historian Stanley Porteus, because Kaua'i was the closest to Japan and most likely to be invaded first by the Japanese, explosives would have destroyed sugar mills, electrical generating plants, harbor facilities, airfields, and any facility beneficial to the attackers. If an invasion came, it would have meant years of postwar economic rebuilding.

Local residents from Hanalei to Waimea stared out to sea at night watching for the light of a Japanese submarine. On December 30, 1941 one broke the surface and fired upon Nawiliwili Harbor, with a dud hitting a fuel tank and damaging a home.

world war II

The Fighting 69th landed in late April to defend the Island, bringing hundreds of regular Army veterans ashore. They camped in Kapaa School and other makeshift quarters. The Battle of Midway, fought in June 1942, twelve hundred miles to the northwest, turned the tide, bringing security and relief that there would be no invasion. Beginning in August 1942 the mission of thousands of mainland troops brought to the Island changed from defensive work to jungle training in Na Pali valleys and other isolated areas to prepare troops for combat in the South Pacific.

The Kauai Volunteers, a local militia, was formed in March 1942 to help defend the Island. Lihue Plantation official Paul Townsley was named commanding officer. Its members were mainly *kama'aina haoles* and Filipinos armed with razor-sharp *bolo* knives. Three volunteer groups were formed: one from Kekaha to Hanapepe, one from Kalaheo to Lihue, and one from Kapa'a to Kilauea. The men supplied their own uniforms and wore a unique Kauai Volunteers insignia on their

The Garden Island

Kauai Under Martial Law

Blackout Rules

Martial Law Orders

Civilian Defense Organization Takes Over This Island

Provisional Police

Keep Calm — Keep the Peace

Instructions to Retail Food Merchants

What to Do in Case of An AIR RAID WARNING

Farmers, Attention

Motorcyclists Wanted

A Personal Message From C. J. Fern

Martial law was declared immediately after the Pearl Harbor attack and lasted through much of the war.

shoulders. All worked regular jobs and trained on Sunday.

Japanese residents including Buddhist priests and Japanese language teachers faced detention at the outbreak of the war. However, Japanese Americans proved their patriotism by working on Sundays in the Kiawe Kommando Korps doing defense work like clearing brush, stringing barbed wire, building trails, and clearing areas for military camps. Young Japanese American men on Kaua'i were the first inductees from Hawai'i in the famed 442nd Regimental Combat Team, with the first volunteers being sworn in on March 11, 1943. A smaller number joined the Military Information Service, serving as Japanese interpreters in the Pacific.

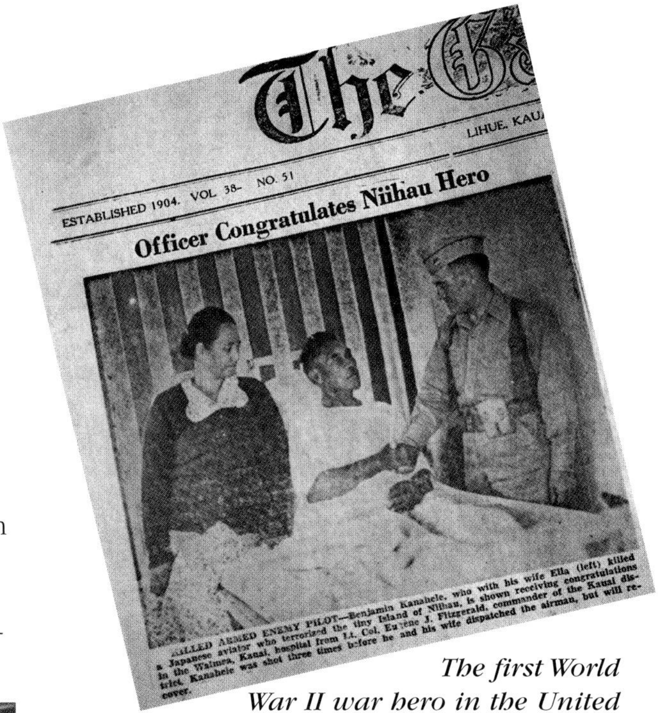

The G

LIHUE, KAUA

ESTABLISHED 1904. VOL 38- NO. 51

Officer Congratulates Niihau Hero

KILLED ARMED ENEMY PILOT—Benjamin Kanahele, who with his wife Ella (left) killed a Japanese aviator who terrorized the tiny Island of Niihau, is shown receiving congratulations in the Waimea, Kauai, hospital from Lt. Col. Eugene J. Fitzgerald, commander of the Kauai district. Kanahele was shot three times before he and his wife dispatched the airman, but will recover.

The first World War II war hero in the United States came from isolated Ni'ihau. Niihau Ranch hand Benekahaka Kanahele and his wife fought off a downed Japanese pilot in the days after the attack on Pearl Harbor, December 7, 1941.

As Army and Marine camps were constructed at Wailua, outside Hanapepe and in other towns, wartime life became a routine. Nighttime brought strict blackouts, with fines for even a glimmer of light showing.

Above Left:
Night-time foxhole duty by members of the 299th Infantry Division including Corporal William Manini, Private First Class Joseph Bukoski, and Private First Class Joe Takata.

Lower Left:
The first Japanese American volunteers in the Territory of Hawaii to be sworn into the army took the oath in Lihu'e and in Hanapepe.

In fear of attacks by Japanese bombers, Lihue Plantation mill was camouflaged during World War II. Hospitals and other large buildings received similar paint jobs. Secret military plans called for destruction of the mills, electric plants, and waterworks if Japanese invasion troops came ashore.

Plantation workers faced wage freezes and weren't allowed to travel to Honolulu without special permission, to make sure they wouldn't stay and take a high-paying defense job at Pearl Harbor. Military rule was instituted, with censorship of *The Garden Island's* "War Daily" and KTOH. All residents except young children were fingerprinted, registered, and given ID cards with wartime serial numbers.

First the economy suffered, then boomed. In 1942 and 1943 there were shortages of some foods, and rationing of gasoline and tires, and liquor. Rice supplies ran low, and meatless days were declared. Victory gardens provided half of the vegetables, with school children helping out by spending one day a week working in fields.

The military population peaked at fifteen thousand GIs and Marines on the Island. Military paydays sparked booming sales at retail stores and movie theaters. Churches filled, bars lit up, and new restaurants opened—nineteen in Kapa'a alone. Young women were employed as air spotters in Lihu'e and taken for joy rides in bombers. Shops at Kalaheo Vocational School, the forerunner to Kauai Community College, turned out machine-gun mounts, and repaired Army equipment. More bank branches opened due to the influx of GI spending. Local craftsmen found a bonanza in making and selling Kaua'i souvenirs. The lack of manpower on the plantations accelerated the mechanization of field work.

Rest and relaxation (R and R) visits for airmen flying missions in the South Pacific left bittersweet memories for families who took them in, with some of the young airmen later killed in battle.

The war also loosened up the tightly structured plantation society as thousands of

The Kauai Volunteers march along a dusty road near Kekaha Plantation.

outsiders brought new ideas, new ways of doing things, and tales of life in Brooklyn, San Diego, and the Midwest. Federal funds flooded in for county projects including the creation of a fire department, plus improvements to roads and harbors. For the first time women worked in plantation mills and pumped gas.

All told, thirteen hundred Kaua'i men went off to war. Forty-eight died, many as members of the 442nd Regimental Combat Team. The 442nd fought in France and Italy, seeing action in pushing the German Army north of the Arno River in Italy, in liberating the town of Bruyeres in France, and in a number of other key battles. The Kaua'i veterans' names are enshrined on a monument in the War Memorial Convention Hall in Lihu'e. After the war, returning GIs came home armed with the GI Bill as a ticket to a future off the plantation. Some arrived with wives from the mainland and foreign nations.

Japanese American veterans came home with a new sense of "can do" that paved the way for major changes that influenced life into the next century, including widespread plantation unionism, wider acceptance of descendants of plantation immigrants in the social world of Kaua'i, and a new political power that ended the Republican domination of local politics and ushered in the Democratic era.

Turk Tokita of Lihu'e fought as a member of the 442nd, receiving two Purple Hearts and a Bronze Star during combat in France and Italy. Tokita took part in the rescue of the "Lost Battalion" of the 36th Division in France, an action considered one of the hardest fought in World War II.

"I never expected to live," said Tokita, looking back to his war years. "I said if I ever get back, I'm going to try to do some-

School children take part in World War II first aid drills.

A World War II era marine camp and Wailua Race Track near the Wailua Golf Course. The large freighter S.S. Andrea F. Luckenberg *went aground on a reef offshore of the camp on March 12, 1951 with no loss of crewman or officers. The ship was later scuttled after salvage attempts failed.*

thing for myself, for my people, for my parents."

During his travels across the mainland before and after his European combat experience, Tokita said he was inspired by getting a look at America, and he came home with a plan to help Kauai's people: use the GI Bill to go to school, return home, become politically active, and work for statehood for Hawai'i. Overall, he saw the necessity of breaking the dependency on a plantation-centered economy as a key in providing new and better work opportunities for Kauai's people.

Tokita studied in Los Angeles and Chicago and had a successful career in Lihu'e as a commercial photographer. In 1950 he became a Young Democrat, joining forces with Honolulu-based 442nd vet Daniel Inouye, and played a key role in gaining a majority of votes for Democratic candidates in the key elections of 1954. Since statehood in 1959 he's worked on the campaign of every Democratic gubernatorial candidate.

"From our war experience we were able to stay together as a tight team; we knew we would have to do that if we were going to change things," Tokita recalled.

"Politics control the economy; if you control the economy you control where people are going to work," he commented on the change brought about in postwar years by political change and diversification of the economy.

CHAPTER 5

the modern era

Kaua‘i faced the end of World War II with big questions about the future. The gold mine of military spending was waning as defense troops pulled out, and how to build on the boost concerned local businessmen.

Life on the still-rural island was in some ways the same as in 1940. Plantations still dominated daily life in most towns, though forced mechanizations meant fewer jobs, and unionization was coming quickly, ending perks like plantation housing.

Population waned as the wartime generation coming of age found their homeland lacking and left for better paying jobs in Honolulu, California, and elsewhere. Others spent their GI Bill tuition money at mainland colleges and relocated permanently following graduation. By 1950 the population slipped to 29,900, down from 35,000 in 1940.

However, the outside world learned about Kauai's beauty through tales told by GIs, and through films made on Kaua‘i. Some soldiers like Spike Roduit of the Fighting 69th from New York married local girls and returned to stay. The connection with the global travel world grew with improved postwar communications, the creation of multiple island tour packages, and thanks to off-island promotions led by Walter Smith, of Smith's Motor Boats at Wailua, and other Kaua‘i businesspeople.

A modern airport opened at Lihu‘e in 1950 on a plain along the coast where stevedores for the port of Ahukini once lived. The construction of a bulk sugar warehouse on a plateau above Nawiliwili Harbor boosted sugar growing.

In 1959 direct jet flights from the West Coast to Honolulu launched the modern-day visitor industry, which by 1980 played the biggest role in the Kaua‘i economy, topping sugar after over a hundred years. By the mid-1960s inter-island jet flight service began, greatly increasing the pace of Kaua‘i visitor arrivals.

Tragedy occurred when a *tsunami* struck in 1946 and 1957, washing out the bridge at Kalihiwai, killing local residents, and devastating coastal neighborhoods at Ha‘ena and Kealia.

The coming of statehood in 1959 brought even bigger changes, and an end to the territorial government formed in 1900. A state government was formed, putting many federally controlled departments into state hands. Statehood was celebrated with hastily

(Continued on page 92)

Rolled up denim blue jeans and acoustic guitars were in style for local cowboy bands.

Kaua'i visitor industry history

Centuries ago, visitors from other Hawaiian Islands looked forward to seeing the spectacular scenery and legendary places of Kaua'i. They crossed the Kaua'i Channel in outrigger canoes and enjoyed acquiring a *limu pahapaha* seaweed *lei* as a souvenir to show they had visited Polihale, and walked a trail to Wahiawa Valley to see the large stone Kaua'i Iki. A rarely seen attraction was the *o'ahi* fire cliff display when burning *papala* and *hau* sticks were launched from a platform on the face of Mount Makana to mark special occasions.

By the 1840s a primitive visitor lodging set up by Kaumualii's last queen, Debora Kapule, stood on the north bank of the Wailua River, near where the main entrance to Coco Palms Resort is today. Debora's home became a favorite overnight rest stop for traveling parties heading to or from Hanalei. The historian and novelist James Jackson Jarves visited Kapule and found a "white style" table with knives and forks, tea with sugar, and enjoyed sleeping on a twelve-foot square bed of *lau hala* mats topped with a *kapa* blanket.

The Kauai Inn evolved from the Lihue Hotel after the aging hotel was modernized. A spacious lawn looking out towards the Ha'upu Range took the place of a beach for guests wishing to sun tan.

During the 1850s makeshift hotels and restaurants opened in homes to accommodate whalers landing at Koloa Landing and Waimea.

From the 1820s up to the 1890s American Protestant missionary families entertained the occasional Western traveler. Mrs. William Harrison Rice captured those days in her journal: "There were neither boarding houses nor hotels on Kauai in that day, and the few white residents kept open house. No one thought of refusing the stranger food and shelter. Whenever a schooner came in no one could tell what travelers would arrive, expecting board and lodging, compensation being left to the guests. Accommodations had to be found for them, often to the inconvenience of the

The Maninibolo Cave in the mid-1950s.

Members of the Royal Court dine during Aloha Week festivities at the Kauai Surf.

family who were turned out of their beds for the wayfarers."

By 1890 the *kama'aina* families could refer unexpected visitors to C. W. Spitz's Fairview Hotel. The Fairview is known as the first proper hotel on Kaua'i and was built at the foot of Rice Street. Visitors staying at the Fairview made their way to Kaua'i by sailing aboard a steamer from the docks at Honolulu to Nawiliwili, where they came ashore in a whale boat. They would find a room for two dollars in 1910 at, "by far the most imposing and pleasantly situated hotel outside Honolulu . . . situated about three-fourths of a mile from the Nawiliwili landing, from which place it is accessible by auto over one of the finest stretches of road in the world."

By 1894 William H. Rice, Jr., was listed as manager. He renamed the property the Lihue Hotel and helped to launch the local visitor industry through organizing promotions and the printing of tourist brochures and maps. Jobs as tour drivers opened up when visitors wanted to see the scenic sights, initally in horse-drawn buggies.

The Kuboyama Hotel and the Seaside Hotel at Nawiliwili, and C. W. Spitz's Seaview at Waimea, opened by 1920, serving "drummers" touting their wares out of Honolulu. The Lihue Hotel's dining room and the Tip Top Restaurant uptown in Lihu'e were the main restaurants on the Island during World War II when young soldiers pulled up in truckloads to eat at the only restaurant on the Island with linen napkins and tablecloths. A room and bath were seven dollars per night. In 1946 the property was sold to Honolulu-based Inter-Island Resorts for $125,000 and its name was changed to the Kauai Inn.

The postwar era marked the beginning of modern tourism. The arrival of regularly

Center:
Pioneering Kaua'i entrepreneur Charles W. Spitz opened the Fairview as the Island's first proper hotel in 1890 in Lihu'e, after starting a store at Nawiliwili in 1886. Spitz later founded Nawiliwili Garage and the first wholesale liquor distributorship on Kaua'i, and invested in pineapple canning and other industries.

Left:
The Lihue Hotel entertained visitors, local residents, and traveling salesmen known as "drummers." The sprawling hotel was run by the Rice family, and was a first for Kaua'i. The hotel was located at the foot of Rice Street.

A torch lighting ceremony marked sundown at the Kauai Surf. Popular beachboy Percy Kinimaka led the festivities.

scheduled jet flights from the U.S. mainland to Honolulu in 1959 brought mass tourism arrivals.

Lyle Guslander's Coco Palms Resort helped lead the way in the new era. Guslander began by purchasing a rustic lodge owned by the Hills family near the site of Kapule's inn. He and his wife Grace Guslander developed a grand resort, enveloping guests in a romantic South Seas environment.

Inter-Island turned the Kauai Inn into a modern resort, adding a swimming pool and a pitch-and-putt golf course. The hotel closed at Christmas time in 1963 after Inter-Island opened its new high-rise Kauai Surf on the beach at Kalapaki. Percy Kinimaka's nightly canoe arrival ceremony by torch light added a Hawaiian touch. The five-story building, and its twin built in the boom era of the 1980s as part of the Westin Kauai, are the tallest on Kaua'i.

NAWILIWILI GARAGE
C. W. SPITZ, Prop.
NAWILIWILI, KAUAI TELEPHONE 104
Automobiles to all Parts of Kauai, all hours, Day and Night
Careful Chauffeurs
Autos and light machinery repaired. Plumbing and gas fittings.
Agents for Fisk and Goodrich Tires and Tubes, Chalmers-Detroit, Hudson & Gramm-Logan Truck.
Agents for the Inter Island Steam Navigation Co., Ltd., at Nawiliwili, Kauai.

AUTOMOBILE RATES

From Lihue to Koloa	$ 5.00	From Lihue to Waimea	$10.00
" " " Lawai	6.00	" " " Kekaha	11.50
" " " Kalaheo	7.00	" " " Barking Sands	17.50
" " " Eleele	8.00	" " " Kapaa	5.00
" " " Makaweli	9.00	" " " Anahola	6.00
" " " Kilauea	10.00	" " " Kalihiwai	12.00
" " " Hanalei	15.00	" " " Haena	20.00

Drummers using Trucks for transportation of sample trunks, $10 and $15 per day and board to driver.
One and one-half fare for round trip.

The 1960s marked the spread of the visitor industry. At Wailua the state-funded development of a marina and parks, and restoration of archaeological sites, set the stage for the development of the Kauai Resort on the south bank of the Wailua River, and to the north the Coconut Marketplace and the stretch of visitor accommodations now known as "Kauai's Coconut Coast."

The Guslanders expanded to Princeville, employing art directors from the film *South Pacific* to design the Hanalei Plantation Hotel on the site of the Birkmyre Estate overlooking Hanalei River. On the South Shore the Kimball family sparked the creation of today's stretch of resorts along the Po'ipu coast by developing the low-key Waiohai resort as an extension of their elegant Halekulani hotel on Waikiki Beach. By 1967 the Sheraton Kauai was opened just west of the Waiohai as the first Kaua'i property to be managed by a national hotel chain.

In the 1960s visitor promotions stepped up to boost visitor counts. Funding was raised on-island for promotions by tourism-related businesses. In the late 1960s funding for a Kaua'i Chapter of the Hawaii

Duke Kahanamoku and his wife Nadine help Grace Guslander celebrate an anniversary at the Coco Palms. The famous Olympic swimmer and Waikiki surfer joined the long list of celebrities who planted coconut trees along the lagoon at the Wailua resort.

Visitors Bureau was provided by the state, and through membership dues. Maile Semitekol served as executive director until 1996. Today Sue Kanoho leads what is now known as the Kaua'i Visitors Bureau.

Above:
In the 1950s, Lyle and Gracie Guslander turned the rustic Hills family's lodge at Wailua into the romantic, South Seas-inspired Coco Palms Resort.

Right:
Hawaiian crafts sellers entertain visitors along Kuhio Highway north of Kapa'a in the early 1950s.

Lower Right:
"Kauai Loves You" promotions inspired by former Mayor Tony Kunimura helped bring visitors back to Kaua'i in 1983 following the devastation of Hurricane 'Iniki. Promoting Kaua'i at a Las Vegas travel trade show are: Nalani Ka'auwai Brun representing the County of Kaua'i, Freckles Smith of Smith's Motor Boats, legislator Billy Fernandes, Governor's Kaua'i representative Mary Thronas, Kauai Visitors Bureau executive director Maile Semitekol, hotel general manager Barry Yap, Hawaiiana scholar Rerioterai Tava, Noboru Yamane of Hawaiian Airlines, long-time Kaua'i promoters Chris Chang and Bob Semitekol.

arranged celebrations that turned into grand events. Thousands gathered at Burns Field at Hanapepe for a huge bonfire built by plantation workers from Grove Farm, McBryde Sugar, Kekaha Sugar, and Olokele Sugar aided by county workers, and fueled with *bagasse* and scrap wood. At nightfall over ten thousand people watched as National Guardsmen from the nearby armory fired off fifty volleys of rifle fire. The following evening a parade in Lihu'e honored Charles Rice and the old guard of the Island who began the march to statehood early in the century. An international program with touches from the cultures of plantation immigrant groups was held at Lihue's Isenberg Field.

The 1960s brought the development of resort destinations along the beach at Po'ipu, at Wailua and Waipouli, at Princeville and Nawiliwili. Thousands of jobs were created, and the economy moved ahead, reflected in the opening of the Lihue Shopping Center in the heart of Lihu'e and at the Eleele Shopping Center. Bowling alleys opened; Dairy Queen and Kentucky

Above:
The Hanalei Fire Department enjoyed an idyllic setting near Hanalei town in 1957. The County of Kauai Fire Department organized during World War II.

Left:
The Kauai Police Department provided music at Lihue Airport in the 1950s. The local police department dates back to the mid-1800s when elected sheriffs patrolled the districts of the Island.

Fried Chicken became the first mainland dining and ice cream franchise chains. Entrepreneurs started up adventurous activities for visitors that brought international attention to the spectacular scenery along Na Pali, most prominently helicopter tours pioneered by Jack Harter and other military-trained helicopter pilots.

As hundreds of young men raised on plantations and in towns marched off to war in Vietnam, the youth revolution hit in the 1960s when young surfers began riding waves at Kauai's ancient surfing spots. By 1970, Kauai's Hanalei Bay was rated among the best surfing waves in the world. The surfing sub-culture was well entrenched on the North Shore and West Side, bringing with it hundreds of surfers from around the world, and a growth in police arrests for marijuana and other illicit drugs. The business side of the surfing world brought in a water sport industry of surfboard builders, and later

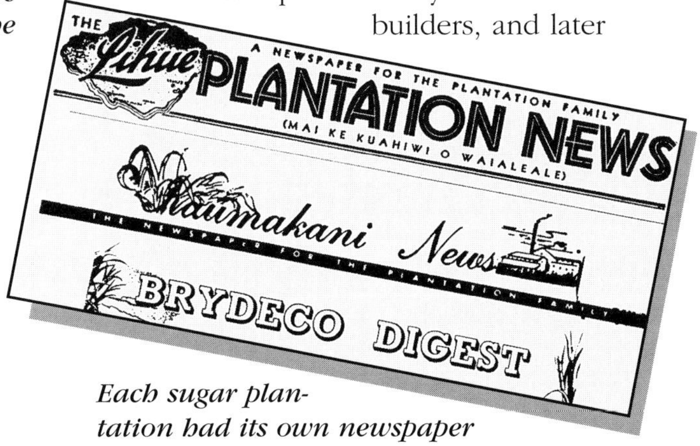

Each sugar plantation had its own newspaper in the 1960s, providing localized news and information about employees, community events, and new sugar technology.

telecommunications

Telephones came early to Kaua'i, with G. N. Wilcox helping to introduce the first lines in 1876, linking missionary Dr. James Smith of Koloa to his home and to Hale Nani, the home of William Hyde Rice in Lihu'e. Wilcox established Kauai Telephone Company, possibly the first incorporated telephone company in Hawai'i. The line was later extended to Hanama'ulu and Kapa'a with eleven subscribers on the line. A separate line was strung between Waimea and Kekaha. By 1900 telephone service was being installed as fast as lines could be built.

In 1899 Marconi invented a wireless system that became known as radio, allowing inter-island and ship-to-shore wireless communication. Sugar planters offered to put up one thousand dollars each to create a territory-wide network, and by 1903 the territorial legislature agreed to subsidize the wireless company for two years, as long as Kaua'i service was added. In the 1910s wireless operators received daily news dispatches for *The Garden Island* in Morse code through earphones, typing out the stories simultaneously.

Early phones rang only once, and often operators spoke in Hawaiian, and puka was used for zero when calls were taken. Party lines were the rule, and repairmen and troubleshooters driving in Model-Ts were often shown aloha by customers and invited in for dinner after fixing problems. The original Kaua'i telephone directory was printed on a single sheet of mimeographed paper.

In the 1920s the organization of workers by the International Brotherhood of Electrical Workers union helped raise wages.

Radio made a big impact in 1926 when the Dempsey v. Tunney fight was broadcast by KGU in Honolulu.

In 1929 Kauai Telephone Company became the last independent outer-island phone company taken over by Mutual Telephone Company of Honolulu. A new central building housed the Lihue exchange, which had gone automatic in 1926. A total of 649 subscribers were online. Mutual was the forerunner to GTE Hawaiian Telephone.

In the 1930s dial-style phones arrived, and in 1931 links to the other islands, the mainland, and Europe through RCA's global system were in place. A wireless telephone station that allowed inter-island radio telephone calls was named Ka Lepa—Signal Hill—and was set on a plateau overlooking Hanama'ulu where native Hawaiians once signaled the arrival of outrigger canoes crossing the Kauai Channel.

The Federal Communications Commission approved a radio transmitter tower for KTOH in Lihu'e and the first official broadcast was aired on May 8, 1940, featuring a live Hawaiian music group led by Albert Nahalea and William Waialeale, Jr. KTOH was run by *The Garden Island* and broadcast from a plantation-style building on Ahukini Road.

On December 7, 1941, military censors took control of radio broadcasts and telephone service, putting censorship in place. By mid-1942 inter-island phone lines were so booked with calls by servicemen that reservations for calls had to be made.

In 1947 Mutual sought its first rate increase since 1922, and reported 2,703 telephones in service. Hotel reservation systems, and office PBX systems, became common in this era. The postwar era also

An early Kauai Telephone Company handset required an operator to make cross island calls.

brought a name change for Mutual in 1954, when the company became Hawaiian Telephone. The 1950s saw phone installations boom, and having a telephone became common in most homes. In January 1955 the Island became the first Hawaiian Island with all-automatic dialing. Intra-island phone charges ended that year.

Television broadcasts from Honolulu were snowy at best on Kaua'i until 1960 when a "satellite" UHF TV station went on the air on the slopes of Mount Kahili. Jack Wada, founder of Jack Wada Electronics, led the effort. Sugar and pineapple companies, local contractors, and other helpers set up the receiver atop 2,800-foot Mount Kahili. An investment of over $25,000 put up by TV dealers paid for the installation, matched by labor and materials from the plantations. KGMB was the first station to install a translator, later joined by KHVH and KHON. The installation proved viewers wanted TV and led to the installation of cable television.

Marshall Derby of Los Angeles brought a modernized cable TV system to Kaua'i in the late 1960s, with a receiver set up at Kukuiolono Park in Kalaheo. In 1970 the Pua Loke subdivision in Lihu'e was wired, and Derby's assistant Bill Gray showed off crystal-clear broadcasts to neighbors gathered around a trailer near the main line in the subdivision. By 1983 most were linked to cable TV, with broadcast signals brought in with a satellite dish. Kauai Cable, Princeville Cable, and Garden Island Cable eventually became Garden Isle Cablevision, making service accessible to most homes.

The telephone service continued to improve, even after set backs occurred during the devastating hurricanes of 1982 and 1992 when thousands of power poles were snapped. A fiber-optic cable was stretched by GTE Hawaiian Tel from leeward O'ahu to a stretch of beach near the Wailua Golf Course. By 1998 all the substations on Kaua'i had digital switching installed.

GTE Hawaiian Tel and AmeriTech began providing widespread cellular service in the 1990s and today most areas of the Island are accessible through the service.

Kauai OnLine, a home-based BBS computer service begun in 1991 by Thayne and Lynne Taylor, grew with the Internet and became Hawaii OnLine in 1994, offering statewide internet access. In 1995 surfing the World Wide Web and sending e-mails across the globe became an everyday event.

Above:
The Sea-Flite hydro-foil ran from a dock near the Aloha Tower in Honolulu to a passenger landing adjacent to Club Jetty at Nawiliwili in 1976. The short-lived cross-channel boat line failed to make a profit and the hydro-foils were shipped to Hong Kong.

windsurfing companies, providing jobs and businesses for newcomers and local craftsmen. Young surfers and hippies built a commune of plastic-tarp tree houses set on land at Ha'ena owned by Howard Taylor, brother of actress Elizabeth Taylor, and at Kalalau Valley and other isolated rural spots. The

This Edition Three Sections 24 PAGES

The Garden Island

LIHUE, KAUAI, HAWAII — WEDNESDAY, MARCH 18, 1959

No Definite Figures on Huge Crowd at Bonfire

Chris Holt and Charles Rice Not Forgotten

Tax Regulations Next Subject of Business Course

Education Leader To Speak Here

Outstanding Workers in Community Chest Cited

Fraga To Speak On Narcotics Here in April

Statehood celebrations held in March 1959 included a huge bonfire at Burns Field.

The Westside community gathers in the 1950s to see the launching of a missile from Bonham air field at Barking Sands. Today Pacific Missile Range Facility is located at the site.

Hurricane 'Iniki devastated the visitor industry in September 1992, destroying homes, businesses, and vacation rentals such as this one at Princeville. Hundreds of rental cars were smashed and vacations of over 10,000 visitors on the Island during the storm were abruptly ended.

anti-work attitude of the hippies shocked hardworking Kaua'i residents.

The 1970s marked the beginning of the decline of sugar growing when C. Brewer's Kilauea Plantation shut down in 1973. Kilauea became a symbol of post-sugar Kaua'i. Predictions had the town folding, but instead residents bought their plantation homes and found jobs at the new Princeville Resort and in Lihu'e, using the modern roads to commute distances that had been unworkable in the past.

The celebration of the bicentennial of Captain Cook's arrival at Waimea held in 1978 might have overshadowed the Island's American bicentennial festivities held on the Fourth of July in 1976 with fireworks and small-town parades.

Visitors flocked to new resorts in the 1980s including a Sheraton-managed hotel overlooking Hanalei Bay, the grand Westin Kauai at Kalapaki Beach near Nawiliwili, and a rebuilt Waiohai hotel at Po'ipu. A collection of world-class golf courses was developed to lure visitors to the resorts, beginning with the Princeville Makai course in the mid-1970s, followed by the Kiahuna course at Po'ipu, Kauai Lagoons at Kalapaki, the Prince course at Princeville, and the Poipu Bay Resort course near the grand Hyatt Regency Kauai resort. Service businesses flourished, supported by work provided by the resorts, and by the spending of thousands of employees. Grove Farm Company's Kukui Grove Shopping Center, with its Long's Drugs and Sears, replaced the venerable Kauai Store, making it possible to buy many items once only available by flying to Honolulu or by mail order.

Kaua'i movie making

Kaua'i is Hollywood's "nearest faraway place" thanks to its majestic scenery, multitude of environments, relatively short flight time from Los Angeles, and skill and diversity of its people. Since 1933 Hollywood casts and crews from the movie capital have traveled to the Island for location filming. Usually, Kaua'i plays an exotic location elsewhere in the world, and its people are cast in a multitude of ethnic roles. Looking back over seven decades of filmmaking Kauai's people can see themselves as South American Indians in *Raiders of the Lost Ark*, French school girls in *South Pacific*, and a Central American island home to dinosaurs in *Jurassic Park*.

The Hollywood-Kaua'i partnership started with location filming of *White Heat* in the summer of 1933. Waimea Sugar Company's three-story mill served as the centerpiece of the interracial love story directed by Lois Weber. Hundreds of extras served as cowboys, servants, hula dancers, and played themselves milling sugar and burning cane fields. Some newly immigrated workers were baffled by what a crew of well-dressed mainland men and women were doing pointing a box at a bunch of workers and actors, running and riding in a cane field.

"It was a real novelty to people, judging by the photographs," says Christine Fayé, whose family owned and ran Waimea Sugar Company. Fayé created an exhibit on the film for the Kaua'i Museum. "Supervisors had the most fun, talking to movie stars dressed to the nines." Fayé says an event like having sophisticated visitors from Hollywood wasn't common in that era.

Following World War II, tales told by returning veterans of the Pacific campaign drove demand on the mainland for realistic location filming of South Seas films. This opened the door to Hollywood coming to Kaua'i.

Pagan Love Song is considered the first major Hollywood feature film made on

Mike Ashman and Eric Knudsen play French colonial roles in Pagan Love Song *filming at Kapa'a in 1950.*

Makua Beach at Ha'ena transformed into mythical Bali Ha'i in 1957 for the filming of South Pacific. *Dozens of extras played French school girls, Solomon Island natives, nuns, and other diverse roles.*

Kaua'i. Esther Williams brought a touch of glamour to the Island and a pattern was set for future productions. The cast and crew stayed in hotels and rented private homes, offered ample meals to extras, and sparked a party atmosphere that went on for weeks. *South Pacific*, filmed in 1957 and using mostly North Shore locations, put Kaua'i on the map in the film world. The towering peak of Mount Makana at Ha'ena became known as Bali-Hai, a mythical headland somewhere in the Solomons. The 1960s and 1970s brought more films including the remake of *King Kong*. Harrison Ford, Steven Spielberg, and George Lucas began the 1980s with location filming for *Raiders of the Lost Ark*, and Spielberg returned just weeks before Hurricane 'Iniki struck in September 1992 to make *Jurassic Park*.

In all, feature filmmakers use Kaua'i as a backdrop, boosting the economy and providing hundreds of jobs, plus adding filmmaking savvy to Kauai's people, most of whom have either been in film, or have a close relative or friend who has been on screen.

Popular 1960s actress Yvette Mimieux sits atop a Cadillac taxi in the parking lot of the Kauai Surf waiting to go on the set of Columbia Pictures production of Diamond Head *in 1962. Co-star James Darren and George Chakiris join Mimieux in watching a local dancer perform.*

"When big movies come to town lots of people work," says Judy Drosd, the Kaua'i film commissioner. "They work in set construction, catering, as drivers, assist in the art department, and many, many people work as extras." She also appreciates the cooperation provided by the people of Kaua'i. "Kaua'i is an island of movie enthusiasts. People love to be both in movies and then go to the movies to pick out their friends on the silver screen. It's more than just the beautiful scenery that lures the filmmakers here; it's a community of people with an enormous diversity of faces, and aloha for the filmmakers."

A young Larry Ramos strums his ukulele for Hollywood filmmakers in 1950. Ramos went on to a career on stage, and in the rock world as a member of the New Christy Minstrels and The Association.

A replica of Honolulu's China Town c. 1900 erected in Lihue Plantation cane fields at Kipu burnt down during the filming of United Artists production of James Michener's The Hawaiians *in the fall of 1969. Extras from Lihu'e and nearby towns peopled the set.*

Kaua'i coffee

In February millions of white coffee blossoms appear in the rolling fields of Alexander & Baldwin's Kauai Coffee Company estate along the highways that stretch from Kalaheo to Numila. By summer the trees are laden with coffee cherries, and by fall the harvest is in, ready to process and bag for shipment across Hawai'i, the U.S. mainland, and overseas.

The successful coffee estate is Hawaii's largest and produces over half of the coffee grown in the Hawaiian Islands. It is the fulfillment of a long-standing quest on Kaua'i, the dream of a series of early American, English, German, and French pioneers in the mid-1800s.

The roots of coffee on Kaua'i, and in Hawai'i, date back to a small plantation in Manoa Valley on O'ahu. There, English horticulturist John Wilkinson planted Brazilian coffee seed in 1825 brought to O'ahu aboard the British warship *H. M. Blonde.* From this small plot, Kaua'i missionary Samuel Ruggles of Waimea obtained the seed that started the Kona coffee industry on the Big Island.

The Manoa plantation probably seeded the first large-scale planting of coffee on Kaua'i, undertaken by Ladd and Company at Koloa Plantation in 1836. Thousands of coffee trees were planted alongside the sugarcane fields of Hawaii's first commercial sugar plantation. The enterprise is located midway between Lihu'e and the present-day Kauai Coffee Company processing plant in Numila. Businessmen back in Honolulu carefully observed to see whether sugar or coffee would be the crop that would give Hawai'i a new industry. While the coffee plantation failed at Koloa, its commercial potential inspired American Charles Titcomb to plant coffee in Hanalei Valley, using coffee seed shipped in from Kona in the early 1840s.

Titcomb later planted coffee at Kilauea, after selling out his Hanalei holdings to

Red ripe coffee cherries in Kauai Coffee Company fields between Numila and Kalaheo.

Robert Crichton Wyllie, the Scot who served as Kamehameha IV's foreign minister and founded Princeville Plantation. Titcomb's competition at Hanalei was his neighbors. In 1842 Frenchman John Bernard and Englishman Godfrey Rhodes leased lands on the east and west sides of Hanalei River and opened the Rhodes and Company Coffee Plantation. Their coffee fields covered seven hundred acres and had hundreds of thousands of trees. German immigrants Wundenberg and Archer grew coffee at Kuna, about two miles up Hanalei Valley from where the Hanalei Bridge crosses the Hanalei River today. Together the coffee plantations of Hanalei exported 21,298 pounds of coffee beans from Hanalei between July 1850 and June 1851. After setbacks due partly to pestilence, by 1853 the fields were consolidated into Wyllie's Princeville Plantation. A young guest of Wyllie's at Hanalei wrote that the coffee trees seemed to be snow covered for a mile along the banks during the springtime blossom.

About twenty miles to the south, English horticulturist Thomas Brown planted coffee at his estate near Wailua Falls in 1846, in combination with a dairy. Brown's wife was Godfrey Rhodes' sister, and the family ties might have influenced his decision to plant coffee on a plateau above the falls. Unfortunately, Brown failed at coffee, too, and moved on. Another coffee pioneer was Hermann Widemann, the founder of Grove Farm Plantation in Lihu'e.

All the early coffee planters were besieged with problems ranging from plant diseases, droughts, and lack of laborers, plus an up-and-down demand for coffee driven by strikes, and losses, in the gold fields of California. The industry died out on Kaua'i by 1857.

A second attempt at launching a coffee industry was tried in the 1890s when Englishman W. E. H. Deverill developed a plantation on leased government lands in Hanakapi'ai Valley. With his wife and young children, Deverill traveled to the valley to plant trees and harvest coffee cherries. The remains of his camp and coffee processing plant can still be seen on hikes to Hanakapi'ai Falls.

Then, for nearly a century, Kaua'i left commercial coffee growing to Kona, although small quantities of coffee cherries continued to be harvested from backyard trees and in the wild for home roasting and drinking. In 1987, Alexander & Baldwin committed to diversifying its McBryde Sugar Company plantation with plantings of coffee trees. Today the Kauai Coffee estate covers thirty-four hundred acres of former sugar-cane fields, providing work for scores of Kaua'i residents, and an export product growing in popularity among coffee lovers around the world. By the mid-1990s the size of Kauai Coffee Company's annual harvests had surpassed those of the entire Kona District.

To handle the bounty of an estate stretching from Koloa to 'Ele'ele, Kauai Coffee Company had to reinvent the way coffee was processed. The company has become a world pioneer in the mechanical harvesting, processing, sorting, and grading of the crop. Combining technology both from other coffee-growing regions and other industries with modern process control, Kauai Coffee Company is today at the cutting edge of the industry, often playing host to coffee people from around the world.

After its fleet of harvesters gathers the coffee cherries, trucks deliver them to the processing plant at Numila, where McBryde Sugar's 1900s-era mill once stood. There, the cherries are separated, "pulped," and washed, and the beans (the seeds newly freed from the cherry pulp) are gently dried for thirty-six to forty-two hours. The resulting "parchment" coffee is then milled and polished, and the beans graded. The whole process is designed not just to produce green coffee ready for roasting, but to sort by quality. In the lab, samples of each batch are roasted, ground, and brewed, then carefully cupped (tasted) for quality. State Department of Agriculture inspectors independently check and certify the grade and origin of the one-hundred-pound burlap bags full of beans, tagging each one.

However, only a small portion of these bags stays on Kaua'i for serving at the Kauai Coffee Visitor Center, adjacent to the processing plant, and for sale at coffee shops and retail outlets not only on the Garden Island itself, but across the state. Most of the crop is shipped to roasters and wholesalers across the U.S. mainland and beyond, bringing the flavor of Kauai's earth, sun, and gentle rain to coffee lovers coast to coast as well as in Japan and Europe.

Kauai-grown coffee beans are fresh roasted at The Kauai Coffee Company's visitor center at Numila. Photographs and video presentations presented at the center tell the history of coffee growing on Kaua'i.

CHAPTER 6

the new millennium

As Kaua‘i enters the new millennium, the roots of its ancient past continue to strengthen the endeavors of its people. Taro growing is flourishing, with new generations opening poi mills to process a food brought to the Island centuries ago. Alongside the age-old agricultural practices, fiber-optic lines strung from one end of Kaua‘i to another are providing the connections needed to thrive in the digital age.

Cutting-edge solar planes are setting global altitude records as they soar over the mountains, plains, and beaches, snapping infrared photographs that are helping coffee growers analyze their plantings. The Pacific Missile Range is on the cutting edge of theater missile research and experimentation and has in place the only testing range in the world that runs from the depths of the ocean to outer space.

The familiar smell of molasses still fills the night air during sugarcane harvests with three plantation mills in operation, continuing an industry begun on Kaua‘i in 1835—a tradition that is now gone in Hawai‘i except for one mill on Maui. Diversifying agriculture is going strong, with coffee, papaya, corn seed, tropical flowers and fruits, and even such diverse crops as asparagus growing in former sugarcane fields.

Ecotourism and cultural tourism are adding a new dimension to grand hotels overlooking what are arguably the world's most beautiful string of beaches. Visitors are choosing to sink their roots on Kaua‘i, purchasing vacation homes and intervals at vacation resorts. Now fully recovered from the devastation of Hurricane ‘Iniki which struck in September 1992, the visitor industry is gearing up for a new century, new challenges, and continued prosperity for its workers.

Filmmakers and photographers continue to find Kaua‘i a spectacular location, enhancing the image of the island across the globe, and providing employment for hundreds each year.

Today, ships docking at Kaua‘i ports include larger versions of Matson's container and roll-on, roll-off barges; twice-weekly inter-island barges carrying Young Brothers and Sea-Land containers and conventional cargoes; the bulk sugar Matson Navigation vessel *Moku Pahu* that carries thirty-four thousand tons of raw sugar to the C and H refinery; a weekly call of the American Hawaiian Cruise ship S.S. *Independence*; and an increasing number of foreign-flagged world cruise ships; plus an assortment of specialized barges carrying fuel oil, gas, sand, propane, concrete, lumber, and construction equipment.

Hawaiian Air and Aloha Airlines fly a full schedule of daily flights to Lihue Airport, which is being improved to serve as the twenty-first-century gateway to Kaua‘i. The largest fleet of tour helicopters in Hawai‘i takes visitors to see the wonders of the Island.

Left: ***NASA astronauts and medical experts traveled to this base at Koke‘e to help guide America's first manned space voyages. Visiting astronauts included Neil Armstrong, the first man to walk on the moon, who manned the tracking station in 1965.***

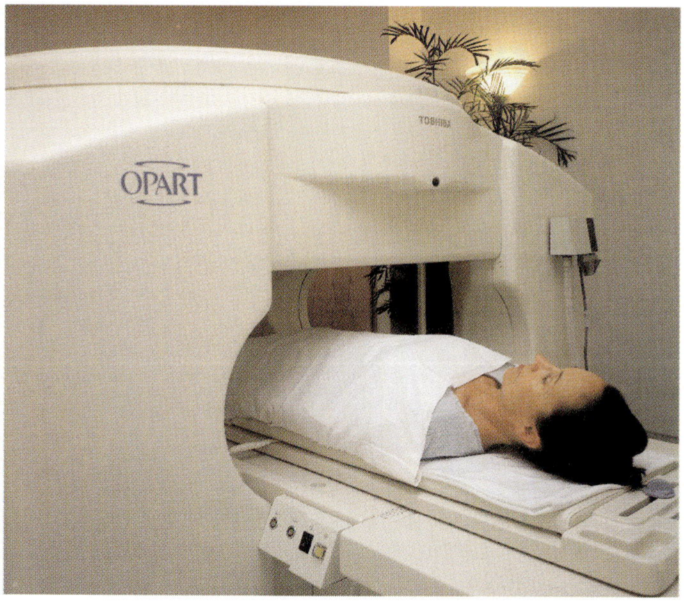

The MRI scanner provides advanced diagnostic services for patients at the Wilcox Health System.

New generations of business people are adapting their family businesses to new times. The Bareback Inn and Koa Trading in Lihu'e, Ching Young Village in Hanalei, the Green Garden restaurant in Hanapepe, and dozens of other businesses built by immigrants and their families are poised for the twenty-first century.

Kauai Builders, Shioi Construction, and a long list of other contractors build the commercial plants needed for new growth and modern homes.

Branches of Honolulu banks continue to serve Kaua'i, building on service that began a century ago. Credit unions first formed in 1938 remain as cooperative community ventures, enabling residents to purchase homes and automobiles, and providing a wealth of financial services.

Federal, state, and county employees make up the largest single sector of the work force, including educators, land conservationists, plant researchers, police and fire services, hospital workers, airport and highway maintenance, water safety and port controlers, and tax collectors.

Native Hawaiian enterprises are growing as the importance of the spirit of aloha in all undertakings is encouraged as a benefit both to island guests and island residents, and a sure way to ensure success.

Guests are entertained by friendly employees at the visitor center for Pahio Resort's Ka Eo Kai property at Princeville. A growing number of visitors are making a long-term commitment to Kaua'i by purchasing intervals at the popular resort.

Advanced farm equipment is used to tend fields of seed corn on the Westside. Novartis, DeKalb, and Pioneer Hi-Bred take advantage of three annual growing seasons on the sunny coast.

Taro grower William Haraguchi inspects taro corms and prepares hulis for planting at his family's farm in Hanalei Valley.

jack harter

When a young man arrived on Kaua'i in 1962 with plans for a helicopter service, the tempo of life was very slow. There were no traffic lights on the island, fields of pineapple still graced the land, three small hotels could accommodate a few visitors, and Hawaiian Airlines still flew old DC3 tail wheel "gooney birds."

Into this subdued, agricultural scene Jack Harter proposed to introduce scenic helicopter tours. He had just completed a contract flying helicopter tours at the Seattle World's Fair. Now he dreamed of starting a copter service on the Garden Island. A wealthy mainland investor was willing to participate, but only if local backing could be found to assure better acceptance. The usual response to his proposal was "go to Maui." He knew that Kaua'i was ideal for copters, and finally found local backing from a resort developer who was building a hotel at Poipu to be called the Waiohai.

The visitor industry was about to grow. Hanalei Plantation Hotel had just opened on a site where scenes from *South Pacific* were filmed. The Kauai Surf Resort was nearing completion at Kalapaki Beach and cruise ships were now making routine visits. Most people still found it difficult to imagine success for a helicopter service planning to charge $120 per flight hour.

Only one suitable copter model was available. The Bell J3-A was considered safe enough for President Eisenhower, and he became the first Chief Executive to use a copter. Unfortunately, the Helicopter of Presidents, as it was called, could only carry three passengers plus pilot. The result was often an empty seat when couples refused to split up.

With $10,000 in working capital, a $65,000 loan to buy the copter, and Jack Harter as the only employee, Kauai Helicopters was airborne. The first scenic tour offered was a round trip flight along the Na Pali coast from a heliport at Dry Caves Beach Park. The price was $20 per person.

From that inauspicious beginning, the industry grew to a peak of fifteen separate helicopter companies flying over thirty copters around the island. Then Hurricane 'Iniki hit in 1992 and blew away the visitor industry for over a year. Fortunately, he had the only helicopter flyable after the hurricane and was in demand for reconstruction. As hotels reopened and visitors began to return, the industry reached ten companies flying over twenty copters.

When Jack Harter is asked what memories of the pioneering years are most vivid, he is quick to answer, "freedom of flight." With only one copter flying, there were no noise complaints, no rules against landing in wilderness places and no pesky FAA around. Honeymoon couples could be left on isolated beaches or at a secret waterfall with guaranteed privacy. It became fashionable for local people to have guests picked up on their front lawns for a scenic flight. Every hotel had a heliport. The U.S. Geological Survey was able to check the Waialeale rain gauge often instead of climbing to the top once a year. Plantation management found the copter perfect for field surveys. Best of all, the fantastic scenery could be used in movies and TV films. The new service even inspired one movie starring Elvis Presley as a copter pilot.

Jack Harter climbs out of Kalalau Valley in the mid-1960s aboard his Bell Helicopter. Na Pali landings permitted in this pioneering era of Kaua'i helicopter touring allowed Harter to drop off honeymooning couples and hikers in remote valleys.

The Marriott Resort & Beach Club at Kalapaki Bay provides oceanfront luxury on the edge of Lihu'e, the modern commercial and political capital of Kaua'i.

Flights to the plateau at Wai'ale'ale to check U.S. Weather Service rain gauges are a tradition for Jack Harter. Here he stands atop Wai'ale'ale in the mid-1960s with one of the original rain gauges, hauled up the mountain early in the century.

More copters were needed as word spread that exploring Kaua'i by air was real adventure. Competition soon arrived when tour companies began including flights in their itineraries. Pioneering had ended. Flight routes and altitudes are now restricted. No landings are allowed.

After nearly forty years and countless flights, how can Jack Harter still enjoy coptering Kaua'i? "Because the island is so incredibly beautiful, and I still haven't seen all her moods," he answers.

Above*:*
Jack Harter drew up flight routes, decided upon pricing, created a logo, and chose attractive photographs for this early black and white Kauai Helicopters brochure.

Producer Howard W. Koch, actor/director Frank Sinatra, and pilot Jack Harter, 35mm camera in hand, on the set of None but the Brave *at Pila'a in May 1964. Harter flew Sinatra from a beachfront home at Wailua to the remote location each day.*

Electrical workers climb a pole high above Nawiliwili to maintain Kauai Electric lines.

Light-filled fiber-optic lines and digital switches link Kaua'i to global markets and promise to be the infrastructure of new high technology businesses.

electricity on kaua'i

As with most nineteenth-century technological advances, electrical service came to Kaua'i homes and businesses years after it was common in mainland cities and Europe. Perhaps the first electricity generated on the Island lighted the home of G. N. Wilcox, generated by a water turbine he set up in a stream near Lihu'e, or perhaps at the home of H. P. Fayé, the West Side plantation owner. It is known that the town of Anahola wasn't electrified completely until the 1950s, and sections of Hanalei Valley had no power lines installed until the late 1960s.

Today Kauai Electric, a subsidiary of Stamford, Connecticut–based Citizens Utilities, is the only electric company on Kaua'i, though its roots go back to a number of electric companies formed in plantation towns in the first half of the twentieth century. First, electric lines gradually snaked out of plantation mills, lighting up offices and mill camps. The first major centralized electrical source was McBryde Sugar Company's twenty-four-hundred-kilowatt Wainiha hydroelectric plant. Wainiha was chosen because the flow of the Hanapepe River wasn't consistent enough to generate a steady flow of power. The plant went online in 1906, sending electricity through mountain passes across a thirty-four-mile power line that stretched to Numila for power milling operations, running elevators that loaded sugar freighters, and to pump irrigation water up through artesian wells. To secure the lease for the plant site and stream use, McBryde sent out Kaua'i-born engineer William Rowell. Negotiations were tough until Wainiha Hui leader Judge Kakina let him know, "No lease, no pig." Once Rowell put on a grand lu'au in Wainiha everything went smoothly. The plant was run by manager Alfred Menefoglio, who lived in a comfortable beach home at Ha'ena, aided by the Gomez brothers from Kalihiwai. Great care was taken in the early days of electricity, for lack of modern safety equipment meant touching a wire could bring instant death. The Wainiha plant was the most powerful one west of the Rockies. Gradually, diesel generators installed outside mills became common, providing the energy for lights in plantation camp homes.

In 1913 Lihue Plantation's electric plant was completed and lit up hundreds of homes from Hanama'ulu to Nawiliwili, pow-

The consolidation of electrical companies run by plantations and entrepreneurs across the Island became Kauai Electric.

Work on remote mauka power lines required horseback rides into the Island's interior by Kauai Electric workers.

ered by a hydro-electric plant along Waiahi Stream. In 1921 the Lihue Plantation mill was electrified. In 1928 Hanapepe got electricity.

By 1929, McBryde Sugar's Kauai Light and Power Company was distributing power for domestic use on the southern, western, and northern sections of the Island, with a report in *The Garden Island* stating, "Old oil lamps [are] giving way to bright electric lights. All the camps from Koloa to Eleele which were dark and gloomy are now or will be brightly illuminated, and outsiders whose homes are near the transmission lines will also have the benefit of electricity." Power to the central and eastern sections was provided by Lihue Plantation (LP) through its subsidiary, Waiahi Electric Company, which was named after a mountain stream that provided hydropower. In 1937 LP installed a two-thousand-kilowatt turbine generator and brought electricity to its operations and plantation camps from Lihu'e to Kealia. Smaller companies like Kapaa Electric Company, Waimea Electric Company, Lihue Ice Company, and other town-based companies generated electricity for a handful of neighborhoods. The construction of military installations during World War II accelerated the growth and consolidation of the electric companies. The subsidiaries were merged early in the 1950s, taking the name of the Kauai Electric Company, which then served a predominantly agricultural economy.

Electrical usage grew rapidly in the 1960s as the visitor industry expanded, and with it, the Island's economy and population. As a result, additional power was needed. The owners of Hawai'i-owned Kauai Electric Company became disinterested in providing the capital for the badly needed generation, transmission, and distribution facilities. Hence, they elected to sell Kauai Electric Company. The owners negotiated the sale to Citizens Utilities Company in mid-1968, and Kauai Electric became a division of Citizens in January 1969.

To supply the power needs arising from the growth of the Island, Citizens has invested more than $180 million (most recently, over $40 million to restore the system due to Hurricane 'Iniki) during the past twenty-six years—enabling Kauai Electric to progress from a simple distribution system to a modern full-service utility. In 1999 Kauai Electric currently serves 29,000 customers, with an installed generation of 110 megawatts, all run by a team of 122 employees.

In the five years prior to Hurricane 'Iniki the use of electricity grew about 7 percent per year until the storm devastated Kauai Electric's system, which is still being restored. Loads in early 1993 dropped some 15 percent compared to early 1992, but returned to pre-'Iniki levels in early 1999.

A Gay & Robinson Company cane haul truck heads to the company's mill at Kaumakani. The rich, red soil of the plantation is among the richest for sugar growing with yields of over fourteen tons per acre.

A new stainless steel vacuum pan installed in 1999 at Gay & Robinson's mill will enhance the company's capability to refine sugar. Company president Alan Kennett leads a tour of the facility.

Restored plantation homes and buildings that make up Waimea Plantation Cottages are preserving the past of Kaua'i, and providing guests with a unique experience.

CHAPTER 7

Lunas who ran the fields for the Hawaiian Sugar Company in the 1920s at Makaweli pose during a break from work. In the center of the photo is photographer Edward Deverill who may have taken the photo by pulling a string attached to the camera shutter.

districts westside

Two young men look across an empty Mana Plain in the 1930s.

The third location of Koloa Mill was the last for the first commercial sugar plantation established in the Hawaiian Islands.

Kekaha Mill in the mid-1920s.

districts
koloa

Resort development along the coast at Koloa flourished in the 1960s. The Waiohai resort was then the centerpiece of what became known as the Poipu Beach Resort.

Fishing sampans tied up at Kukui'ula Harbor in the 1920s in an era when most South Shore residents lived in Koloa town.

Lawa'i-Kai Valley planted in rice. Prior to its purchase by Robert Allerton, the valley provided agricultural lands for McBryde Sugar Company workers.

Split-rail fences and dirt roads ran between Lawa'i and Koloa and other rural towns in the era between 1890 and 1920 when trails and carriage roads were widened in preparation for paving.

GEORGE MUNDON. L. K. KAUMUALII.

GEORGE MUNDON & CO.,

—DEALERS IN—

GENERAL MERCHANDISE,

—INCLUDING—

Dry Goods, Clothing, Boots and Shoes, Groceries, Hardware, Crockery, and Fancy Goods.

A First-Class Billiard Saloon Attached to the Store.

Koloa Provision Store,

E. STREHZ, Proprietor,

—DEALER IN—

Groceries, Hardware, Tinware,

CROCKERY, TOBACCO and CIGARS, STATIONERY,

DRUGS AND MEDICINES, CONFECTIONERY, ETC.

KOLOA, ISLAND OF KAUAI.

Post-Office, Telephone Office, and Collection of Customs.

TRAVELERS, Stop at the

Koloa Restaurant

H. APOI, Proprietor,

KOLOA, - KAUAI,

And Get a SQUARE MEAL.

APOI also deals in **Clothing, Dry Goods, Groceries, Etc., Etc.**

General stores provided a wide variety of goods in most towns in the 1880s.

Nawiliwili grew from a coastal Hawaiian village into the Island's main port.

A hukilau at Nawiliwili in the 1930s.

The Wilcox beach home at Papalinahoa once stood near where the main Hale Kauai store is today.

districts
eastside

A crowd of Kapa'a residents watch movie makers film Pagan Love Song *in 1950. The Crossroads cafe was a popular eating spot for Roxy Theater moviegoers and Hawaiian Canneries pineapple workers.*

Huts constructed by immigrant sugar workers provided housing along Koʻolau Road near Moloaʻa Beach into the early 1900s.

A horse racing track, groves of mature coconut trees planted from Samoan coconuts years earlier by German immigrants, and a new airstrip, brought changes to the coast at Waipouli in the 1920s.

districts
hanalei

The Princeville Ranch and rice fields dominated the landscape of the North Shore towns prior to the 1960s.

KAUAI. KAUAI.

THE KILAUEA STORE,

Sydney House, - - KILAUEA,

WHOLESALE AND RETAIL.

Is the Largest Store on Kauai.

—THE STOCK OF—

GENERAL MERCHANDISE

Is the **BEST and CHEAPEST**, Comprising

Wearing Apparel of every Description,
Dry Goods, Hardware, Provisions,
Groceries and Canned Goods,
Saddlery, Books, Stationery,
Toys, Sewing Machines,

Tobacco, Cigars, Cigarettes,

DRUGS and PERFUMERY.

New Goods Received by Every Steamer.

As the **Kilauea Store** is the only one in the Hawaiian Islands conducted entirely on **Cash** Principles, it is able to Retail Goods of the BEST QUALITY at exceptionally **Low Prices.**

The Kilauea Restaurant,

OPEN FROM 5 A. M. TO 8 P. M.

Meals at all Hours.

TEA, COFFEE, and ICED DRINKS.

This Establishment seeks to provide travelers and the public with all kinds of **REFRESHMENTS** cooked and served in the best possible manner. No pains are spared to render this Restaurant second to none in the Kingdom. It is situated half-way between **KAPAA and HAUALEI, at SYDNEY HOUSE, KILAUEA.**

KAUAI. KAUAI.

A newly-built Hanalei Bridge, erected in the 1860s by Princeville Plantation's founder Robert Crichton Wyllie, crossed the Hanalei River upstream from the Princeville Plantation sugar mill.

Lumaha'i Valley, once lined with taro lo'i and populated by native Hawaiians, is today the main Kaua'i land holding of the Bishop Estate.

A village of stores and homes lined the coast at Kalihiwai prior to the arrival of a destructive tsunami in 1957. The tsunami took out the village's coastal bridge and led to the construction of the long, sweeping road bridge that now crosses Kalihiwai Valley mauka of the coast.

CHAPTER 8

Kaua'i Chamber of Commerce President Laurie Yoshida, Senator Daniel Inouye, and Chamber Board Chairman Bob Mullins of Textron join in the festivities at the opening of the new West Kauai Technology and Visitor Center at Waimea in April 1999.

chamber of commerce history

The work of the Kaua'i Chamber of Commerce since its inception in 1913 has been the betterment of the Island's economy, its businesses, and its people. The Chamber continues to uphold traditions built up over nine decades, lobbying for Kaua'i bills in the Legislature, working to promote the visitor industry and Kaua'i as a place to do business, and working hand-in-hand with the almost five hundred businesses, large and small, who make up the Kaua'i Chamber of Commerce.

The first stirrings of interest in forming a Chamber for Kaua'i came in 1911 when automobile garage, liquor distributor, and hotelier C. W. Spitz attempted to gather businessmen together. *The Garden Island* reported: "A move is on foot to organize a chamber of commerce; the idea being supported by a majority of our prominent townsmen."

In the summer of 1913 leading businessmen were approached to organize a board of trade, with "business firms of every class" as members. Manager H. Rohrig of the Lihue Store with Lihu'e businessmen William Hyde Rice, Jr., and Spitz traveled to Waimea and found support.

The Chamber met as an organization in the County Building in 1914, with a Mr. Putnam as the first president. Pressing issues included development of a pier and deep-water harbor at Nawiliwili, "village improvements" for Lihu'e, censorship of the movies coming to the Island, and lobbying for a $100,000 bond for the county to speed up the completion of the belt road.

Meetings alternated between Waimea and Lihu'e. Early committees included one to help homesteaders in Kapa'a, Kalaheo, Lawa'i, and other growing towns. A harbors committee and a postmaster committee were formed. Joining in on territorial-wide tourism promotions began in 1917 when Kaua'i was allowed to join the Hawaii Promotion Committee. Chamber presidents then included E. H. W. Broadbent and Judge Lyle Dickey.

In the 1920s meetings were held at the popular Tip Top Cafe and the Lihue Hotel. Problems to be solved included developing beach lots at Wailua, and speeding up mail service to the Waimea. Philip Rice represented Kaua'i at the Territorial Chamber of Commerce, and William Henry Rice, owner of the Lihue Hotel, was placed on the Hawaii Tourism Board. By 1924, during the presidency of Wainiha power plant manager Alfred Menefoglio, membership rose to fifty. In 1929, J. T. Moir Jr. of Koloa Plantation led discussions on taxing gasoline sales and allowing women on juries.

The opening of the dock and harbor at Nawiliwili in 1930 was cause for a major Chamber of Commerce led celebration. Harbor supporter G.

N. Wilcox became the first honorary member of the Kauai Chamber of Commerce. The election of officers was moved from mid-year to the beginning of year during Moir's reign, conforming to practices of other Chambers in Hawai'i. In 1931 a new constitution was adopted allowing firms, partnerships, and corporations to be eligible for membership, with annual fees ranging from $10 to $100. The new rules brought in support from sugar and pineapple plantations that would be crucial to the future of the Chamber.

In 1935 the first Kauai County Fair was backed by the Chamber of Commerce. At the 25th anniversary celebration held in 1939 members waxed nostalgic, recalling the days of schooner travel to Honolulu, and proudly showing that the size of the Kaua'i sugar crop had doubled over the years. Old timer Eric Knudsen recalled days of automobile travel before all roads were paved.

The coming of World War II in 1941 temporarily slowed down Chamber efforts as the Office of Civilian Defense prepared the Island for possible Japanese invasion. By 1942 Chamber president B. D. Pratt was helping organize entertainment for servicemen. Chamber meetings featured speeches by officers leading the military effort. Assistance for the Kauai Volunteers, and updating county road rules due to the massive increase of traffic due to the war, were on meeting agendas.

The Junior Chamber of Commerce formed in January 1945 under the direction of Allan Ezell, with Toshi Serizawa as a director. Post-war issues for the Chamber included improving schools and looking at steep freight cost increases.

In 1947 construction of a new airport at Lihu'e was a hot issue. The Territorial Department of Public Works wanted to build only one runway, while the Chamber battled for a "complete airport" with a connecting airstrip added to the work to allow flights on Kona wind days. In 1949 flood relief for Waimea was supported by the Chamber, following a devastating flood that hurt businesses and homeowners. Keeping Port Allen open, even as Nawiliwili Harbor grew into Kauai's main port, was supported.

Jack Sheehan of Koloa, president in 1952, asked members to be on the guard for subversive acts of Communists, reflecting the Red-Scare crackdowns of that era. The 1954 meeting saw a call for hiring a paid executive secretary to run a Chamber of Commerce office and to implement a growing number of economic development projects. In 1958 merchants and TV dealers were asked to help subsidize equipment needed to pick up television broadcasts from Honolulu through the purchase of TV advertising. Statehood brought increased tourism and the raising of funds for tourism promotions. As hotels opened and resort destinations formed in the 1960s visitor industry members became more prominent.

The Kauai Dollar promotion was a hit in 1972, with Tad Miura receiving the first shiny dollar. Worries about the brain drain of bright students from Kaua'i were concerns.

Through the 1980s and the 1990s the Chamber continued to grow with the Island during the boom years, and lead post-hurricane recoveries following Hurricane 'Iwa in 1982 and 'Iniki in 1992. Today issues common throughout the Chamber's 85-year history continue to bring together business people: expansion of the runway at Lihu'e Airport, rising taxes, and promotion of the Island as a great place to live, visit, and do business.

KAUAʻI

Chamber of Commerce

bibliography

A Select Kaua'i and Ni'ihau Bibliography

Alexander, Arthur Chambers. *Koloa Plantatiuon, 1835-1935; A History of the Oldest Hawaiian Sugar Plantation.* Honolulu Star-Bulletin Press, Honolulu, T. H., 1937. Reprinted by Kaua'i Historical Society in 1985.

Bennett, Wendall Clark. *Archaeology of Kaua'i.* Bishop Museum Press, Honolulu, Hawai'i, 1931.

Blay, Chuck. *Geology of Kaua'i.* Kaua'i, TEOK, 1998.

Clark, John R. K. *Beaches of Kaua'i and Ni'ihau.* University of Hawai'i Press. Honolulu, 1990.

Cook, Chris. Douglas Peebles, photography. *From the Skies of Paradise—Kaua'i.* Mutual Publishing, Honolulu, 1991.

Cook, Chris, editor. *The Garden Island Guide to Kaua'i.* Lihu'e: Kauai Publishing Co., 1998.

Cook, Chris. David Boynton, photography. *The Kaua'i Movie Book.* Mutual Publishing, Honolulu, 1996.

Cook, James. *A Voyage to the Pacific Ocean...for making discoveries in the Northern Hemisphere in...the* Resolution *and the* Discovery*...vols. 1-3 and atlas.* British Admiralty, London, 1784.

Damon, Ethel. *Koamalu.* Two volumes, privately published. Honolulu, 1931.

Daws, Gavan. *Shoal of Time: A History of the Hawaiian Islands.* University Press of Hawai'i, Honolulu, 1974.

Fayé, Christine. *Guide to Waimea.* Kaua'i Historical Society, 1997.

Joesting, Edward. *Kaua'i: The Separate Kingdom.* University of Hawai'i Press, Honolulu, Kaua'i Museum Association, Lihu'e, 1984.

Kaua'i Historical Society. *The Kaua'i Papers.* Kaua'i Historical Society Publication, Lihu'e, Kaua'i, 1991.

Krauss, Bob. *Grove Farm Plantation—The Biography of a Hawaiian Sugar Plantation.* Palo Alto, CA., 1984.

Rice, William Hyde. *Hawaiian Legends.* Photographs by Boone Morrison, Bishop Museum Special Publication 63, Honolulu, 1977.

Riznik, Barnes. *Wai‘oli Mission House, Hanalei, Kaua‘i.* Grove Farm Homestead and Wai‘oli Mission House, Lihu‘e, Kaua‘i, 1987.

Valier, Kathy. *On the Na Pali Coast—A Guide for Hikers and Boaters.* University of Hawai‘i Press, Honolulu, 1988.

Wichman, Frederick B. *Kaua‘i Tales.* With illustrations by Christine Fayé, Bamboo Ridge Press, Honolulu Hawai‘i, 1985. Two additional books in the series are available.

Wilcox, Carol. *The Kaua‘i Album.* Photographs by John Wehrheim and T. K. Kunichika. Kaua‘i Historical Society, Lihu‘e, 1981.

Ni‘ihau

Gay, Lawrence. *Kainoahou Tales of the Forbidden Island of Ni‘ihau.* Topgallant Pub. Co., Honolulu, 1981.

Moriarty, Linda. *Ni‘ihau Shell Lei.* University of Hawai‘i Press, Honolulu, 1986.

Tabrah, Ruth M. *Ni‘ihau—The Last Island.* Press Pacifica, Honolulu, 1987.

Tava, Rerioterai & Keale Sr., Moses K. *Ni‘ihau—The Traditions of an Hawaiian Island.* Mutual Publishing, Honolulu, 1989.

von Holt, Elizabeth Knudsen. *Stories of Long Ago.* Honolulu: Star-Bulletin Printing Co., 1968.

Index

I

J

K

T

U

V

W

Outrigger canoe paddlers wait to greet the Matson Liner Lureline during a stopover at Nawiliwili. The popular steamship carried passengers on the Hawai'i-California route.

Y

photo credits

David Alexander: Kekaha Sugar contract, letter to Kilauea Plantation, passengers at Nawiliwili dock, ship at Nawiliwili 1930, Filipino house at Hanalei, Lihue Plantation camoflauged, Polihale lookout, hukilau at Nawiliwili, Papalinahoa home.

Mike Ashman: Hawaiian Canneries aerial, Carmen Miranda arriving, KTOH letterhead, Filipino music group, Charlie Fern, Marine Camp at Wailua, band around KTOH microphone, KPD band, Kapa'a street scene, *Pagan Love Song* costumes, Larry Ramos, Hawaiian souvenir sellers.

David Boyton: Wai'ale'ale, makaloa mat, Awa'awapuhi, Alekoko Fishpond, maile-mokihana lei, poi pounders, Menehune ditch, Pu'u-Ka-Pele, Waiehu Falls, Captain Cook statue, Russian Fort aerial, time share guests, corn seed driver, Kalapaki resort aerial, fiber-optic line, Waimea Plantation Cottages aerial, Gay & Robertson cane haul truck, Kauai Coffee visitor center, coffee cherries, Waimea Technology and Visitor Center, John Ena mural (courtesy of Alexander & Baldwin).

Chris Cook: Lihue Soda Co. bottle, tunnel characters, moku map, district map, Waimea graves, Waimea Dairy bottle and container, antique telephone, Prince Kuhio monument, C. Hofgaard, hurricane damage, taro field, Alan Kennett.

Chris Deverill: Lunas at Makaweli, cattle boat.

Nick Galante: Pathfinder over Na Pali.

Gay & Robinson: Immigrant workers in cane train.

Hanalei Museum Collection: Koke'e Stream, Na Pali Hale, lau hala weaver, Maniniholo Cave, Captain Cook postcard, Birkmyre estate, Princeville Plantation, Hanama'ulu mill, Kapaa Sugar Co., Lihue Plantation mill, sugar workers cutting and loading cane, Hanama'ulu train trestle, cane train in plantation camp, Hanalei rice fields, Japanese immigrants, Makaweli road, scow at Lumaha'i, touring car, Tip Top Building, Nawiliwili street scenes, Lihue Library, Bank of Hawaii, Lihue aerial, Hanalei Fire Department, Ko'olau immigrant huts, Hanalei Bridge, Lumaha'i Valley, Kalihiwai Valley, Lihue Hotel.

Jack and Bev Harter: Helicopter over Na Pali, Kauai Helicopters brochure, Frank Sinatra and Howard Koch.

Hawai'i State Archives: Nawiliwili post card, Waimea Harbor map, Hanalei Harbor map, hula ladies, geology map, Barking Sands, Makee Sugar Co. mill, Princeville Plantation mill, Numila, Lihue Plantation mill, Kilauea Plantation aerial delivery, mauka railroad trestle, water buffalo, boat in Nawiliwili Harbor, early Port Allen, schooner and steamer at Ahukini, Waimea Pier, Portuguese immigrants, Filipino YMCA club, road building on North Shore, Lihue Plantation crossroads, Waimea roads in 1800s and 1930s, Inter-Island Airlines at Burns Field, bi-planes at Burns Field, Kalalau Valley aerial, view of southwest Kaua'i, Koloa Mill, Koloa aerial, Lawai-Kai, Lawai Road, aerial view of Lihue, Kapa'a fire, Kilauea Plantation train.

Kauai Electric: Men on horseback, group throwing electrical switch, linemen above Nawiliwili, hydrofoil.

Kauai Historical Society: Swearing in of Japanese Americans in WWII, night time fox hole, first aid training WWII, Kauai Volunteers marching at Kekaha, Duke Kahanamoku and Grace Guslander (Guslander Collection).

Kaua'i Visitors Bureau: Car at Maniniholo Cave, royal court at Kauai Surf, Coco Palms lagoon.

Kawakami Family: Kawakami Store.

Kikiaola Land Co.: G. N. Wilcox, Knudsen family, Westside family picnic, Fayé plantation managers, H. P. Fayé, Leong Pah On, Chinese workers.

Photographer W. J. Senda on the set of White Heat *at Waimea in 1933.*

Pa'a Studios: *South Pacific* filming (W. J. Senda), coffee processing, Captain Cook chanter, Russian Fort drawing, George Kaumuali'i, canoe at Kalapaki, whaling at Hanalei, Dr. Smith, missionary map, guano logo, William Hyde Rice, Rice Ranch poster, public school children, Ko'olau, moving rocks with mules, ship-

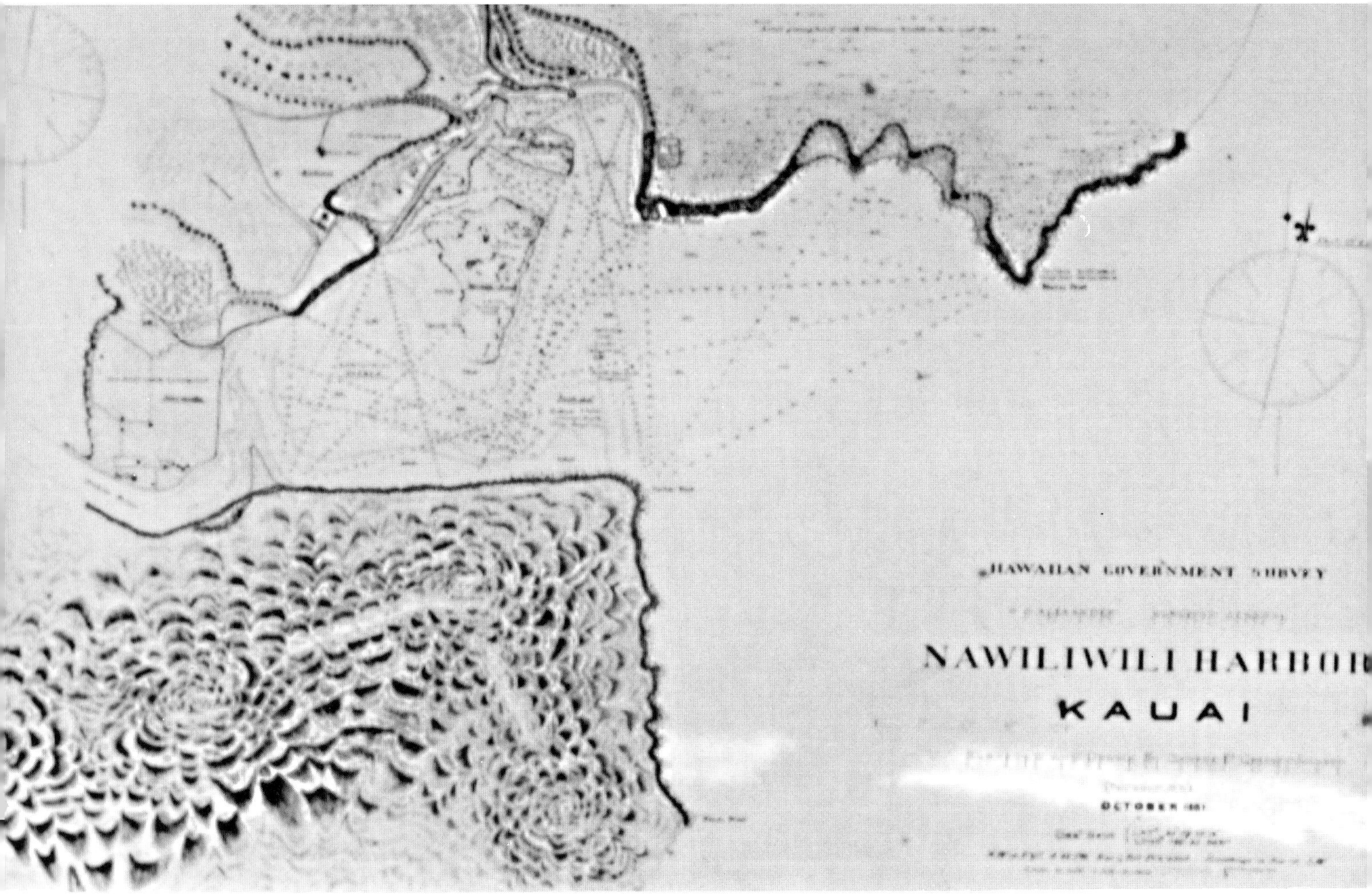

ping broadsheet, Inter-Island logo, rowing ship passengers, Chinese map, Chinese worker, Ha'ena Road (W. J. Senda), fording river on horseback, swimming platform at Kilauea, 1906 Board of Supervisors, Anahola aerial, Aliomanu pineapple fields postcard, Pono pineapple label, Territorial County of Kauai logo, sugar plantation newspapers, PMRF rocket launching, Waimea Port aerial, Kekaha Plantation aerial, Koloa Store poster, Waipouli aerial, Makee Sugar Co. aerial, Kapa'a town aerial, Hanalei-Princeville aerial, Kilauea store advertisement, Bali Hai movie cast (W. J. Senda), stars at Kauai Surf (W. J. Senda), *The Hawaiians* scene, Kauai Inn postcard, C. W. Spitz, Nawiliwili Garage advertisement, W. J. Senda with movie starlet, Kaua'i books, Paul Isenberg, Port Allen aerial, Kaua'i tourism brochure.

The Garden Island Collection: Salt pans, Lihue Soda Co. advertisement, Filipino Businessmen's boxing card, Inter-Island Airlines advertisement, Garden Island Motors logo, Kauai Stores logo, Kanahele headline, martial law headline, Statehood headline, Po'ipu Resort aerial, Nawiliwili aerial, Lihu'e in mid-1960s, Kauai Surf torch lighting.

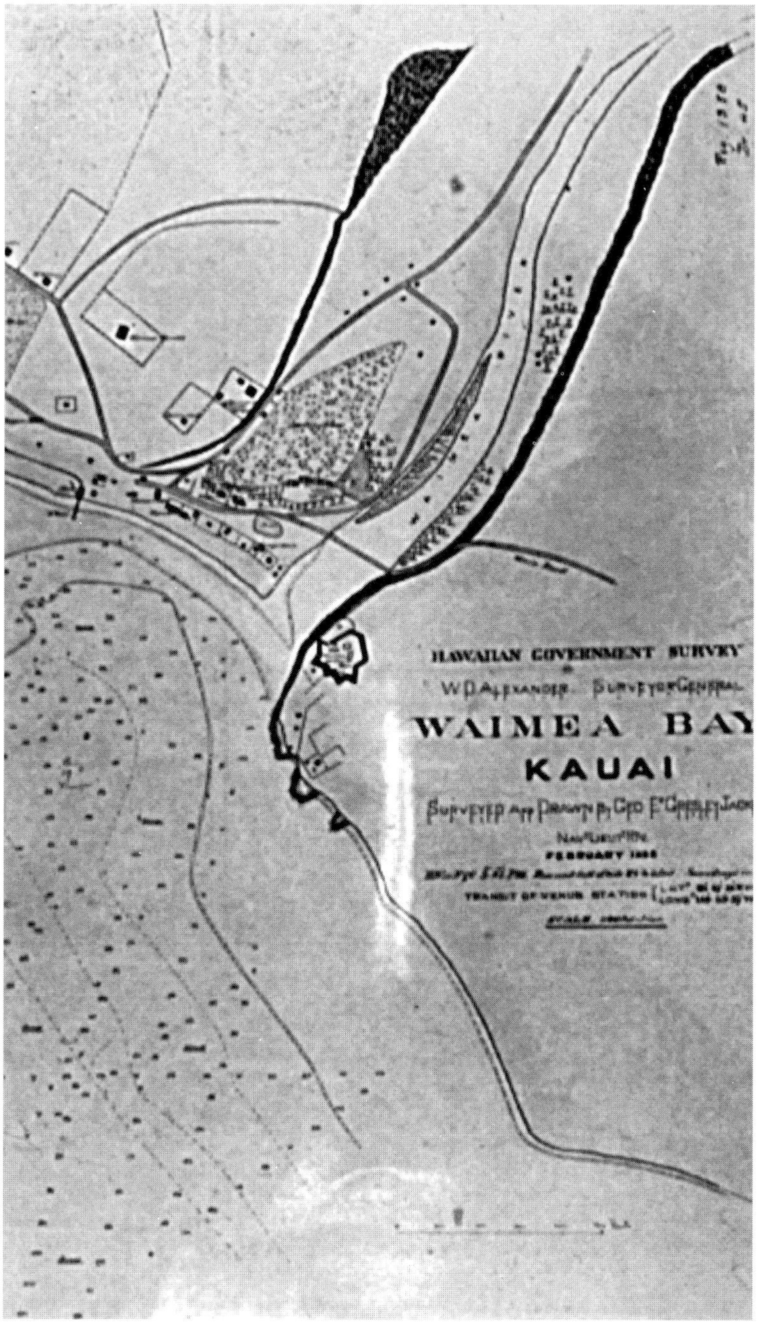

Maile Semitekol: Tourism promotion in Las Vegas.

Ship Store Galleries/Raymond Massey: Detail from Cook arrival sketch.

Harvey Tam: Pineapple cannery workers, Hawaiian Airlines airplane at Lihue Airport, Jack Harter atop Wai'ale'ale, Kilauea Lighthouse cleaners, Lurline at Nawiliwili, Honopu rock with names, NASA radar dome, pineapple pickers.

Wilcox Health System: MRI system.

Wilcox Hospital Foundation: twin babies at hospital, G. N. Wilcox Memorial Hospital.

about the author

about the author

Chris Cook lives and works on Kaua'i. He is the New Media Manager for *The Garden Island* newspaper. Cook has worked for the county of Kaua'i as a tourism specialist, for Princeville Corporation as Manager of Corporate Communication, and has been a reporter and business editor for *The Garden Island.* His books about Kaua'i include *The Kaua'i Movie Book*, *From The Skies of Paradise—Kaua'i*, and *The Garden Island Guide to Kaua'i.* Cook is a graduate of the University of Hawaii at Manoa, and the owner of Pa'a Studio, a small business on Kaua'i.

about the photographer

Photographer David Boynton has been active in conservation efforts and as an educator on Kaua'i for over twenty years. His photographs have appeared in dozens of books and magazines, focusing primarily on native ecosystems and Island landscapes. Currently, he is the environmental education resource teacher for Kauai's public schools, based at the Koke'e Discovery Center.

PAHIO
RESORTS

Lumahai Beach

Much of the film version of the hit Broadway musical South Pacific, starring Rozzano Brazzi and Mitzi Gaynor, was shot in and around Hanalei. Lumahai Beach is Nurses Beach where Nellie Forbush "washed that man right out of her hair."

NaPali Coast

(below)

The NaPali towers 4,000 feet above the pounding surf. The Kalalau trail follows the rugged coastline and is a challenging, as well as, phenomenal two-day hike.

Nothing is as you expected it. The mountains are higher and greener. The sky is more blue and the water crystal emerald. Wild orchids splash color where you least expect to see it. And always the sun is warm and lazy. For most, a special relationship begins. You'll find yourself at one with the island as the caress of a soft breeze recalls memories you never knew you had. You'll feel a part of the heritage of Kauai without knowing why. You'll come to believe that it really is magic. And you'll never feel quite as much at home anywhere else, again.

From The Moment You Arrive Kauai Begins To Work Its Magic

Hearts and imaginations have been captivated by the magic of Kauai for generations. It's the land. The heritage. And most of all, the Aloha of its people. Here you will step back in time and witness the same pristine beauty that enchanted early Hawaiians . . . the bold mist shrouded mountains, cascading waterfalls, dazzling rainbows . . . the spectacular pristine coastline and secluded white sand beaches.

Kauai is a place for romance, a paradise where time stops and stress melts away. Where hurry slows to a delightful crawl and worry is hushed by the gentle trade winds. On Kauai a smile is always in style and nature's beauty becomes more magnificent with the passing of time.

Kauai Rivals The Most Exquisite Vision Of Paradise

By most estimates it took nearly two million years for the boiling volcano (named Olokeke) that formed Kauai to be carved and contoured by crashing surf, winds, and rains into the spectacular cliffs, canyons, valleys and plateaus that now enchant visitors from around the world.

It is Kauai's ancient tradition of pride in its perfection that has inspired the design and service integrity of PAHIO Resorts.

The first people to set foot on Kauai more than a 1500 years ago were daring Polynesian adventurers who set out in 80-foot double hulled canoes in search of their culture's vision of paradise. Using their knowledge of the skies, ocean currents, and the flight of birds, these explorers prayed that their gods would guide them to a new and ultimate place of perfection.

Legend has it that they landed on the eastern shore of Kauai, at the mouth of the Wailua River. The remains of ancient shrines (heiau) found there are a testament to the area's rich historical and religious significance.

As Hawaii's most magnificent island, Kauai radiates an abundance of scenic splendor. It's isolation from the other Hawaiian islands has allowed the evolution of an ecosystem unlike any other. As the home to more than 2500 plants that grow nowhere else in the world . . . it is fitting that Kauai be called the Garden Island.

But Kauai is distinguished from the other islands by more than its lush beauty. Kauai is the reputed first home of Pele, the revered fire goddess whose reign continues today on the Big Island of Hawaii. Here too is the habitat of the Menehune, a race of little people that folklore credits with the creation of some of the island's most spectacular natural features and greatest mischief.

Kauai was the first Hawaiian island discovered by Captain Cook. It is known as the home of Hawaii's first sugar plantation. And, it is the birthplace of the traditional hula.

Kauai Will Forever Define Your Image Of Romance And Adventure

Kauai is the perfect setting for a great variety of tropical adventures. Kauai's beaches are among the most beautiful in the world, and the ocean is a comfortable 75 degrees year round. The warm and clear

It is understandable why the people of Kauai retain a passion for the protection of their perfect island paradise. As the oldest and most remote of the major Hawaiian islands, Kauai is acknowledged as the island with the richest history, the proudest people, and the most intriguing folklore . . . It is the birthplace of the traditional hula.

waters invite everything from swimming and snorkeling to surfing and fishing.

Windsurfing is especially

popular with PAHIO owners. The best place to learn to windsurf is found only a few minutes from Princeville at Anini Beach. The wind and waves are very predictable and perfect for the beginner. It is said that the sport can be mastered in an afternoon.

Surfing remains one of the most popular sports in the islands. While body surfing and boogie boarding are the easiest way to ride a wave, learning to maneuver a surfboard is an irresistible challenge for many visitors.

The clear lagoons surrounding Kauai provide a fascinating view into another world for snorkelers and divers. The coral reefs are home to a great variety of colorful tropical fish.

Kayaking is the perfect ocean sport for almost everyone. In the calm waters of a bay or on the Wailua and Hanalei rivers, you'll be able to sightsee, paddle in the sun, and exercise at the same time.

Cruising the coast on a sailboat or catamaran is a great adventure. Boat tours are the easiest way to experience the rugged NaPali and a must when the humpback whales make their annual visit to Kauai. The island is also renowned for great sportfishing. Few thrills in sportfishing can match the experience of hooking a blue marlin from a charter.

If you want to try something really different, why not explore Kauai on the back of a horse with a paniolo (Hawaiian cowboy) guide, or on the seat of a Harley-Davidson motorcycle with the wind in your face.

On Kauai you will experience a lifetime of memories. Whether you enjoy the exhilaration of an exotic island adventure or the simple pleasure of a romantic walk on a secluded beach. Kauai is a paradise where life is full of excitement, and each

moment is to be savored.

Hollywood Discovers The Splendor Of Kauai

Since Hollywood discovered Kauai in the 1930's, dozens of sites around the island have been used by film directors to create the image of a lush Polynesian paradise. More than 200 films have featured Kauai's stunning beauty.

The 1961 film *Blue Hawaii* starring Elvis Presley, was filmed at Wailua. Television's Fantasy Island opened with a shot of Wailua Falls. The remake of the film classic King Kong was filmed in a remote valley along the NaPali coast. Honapu Valley was transformed into the South American jungle for *Raiders of the Lost Ark*.

George C. Scott sailed down Wailua river in the 1976 film adaptation of classic Hemingway novel, *Islands in the Stream*. In the 1983 production of *The Thornbirds*, Hanapepe was chosen for its similarity to a nineteenth century Australian frontier town.

Scores of other movies and television productions have been filmed on the island including *Jurassic Park* and its sequel, *Honeymoon in Las Vegas*, *Lord of the Flies*, *None but the Brave*, *Uncommon Valor*, *The Hawaiians*, *Lt. Robinson Crusoe, U.S.N.*, *Throw Momma From the Train*, *The Wackiest Ship in the Army*, *Hook*, *Donovan's Reef*, *Diamond Head*, *Body Heat*, and the pilot of *Gilligan's Island*.

1. Sailing on Hanalei Bay 2. The lighthouse at Kilauea 3. Couple preparing to surf 4. Sailing the open sea 5. Hanalei Valley 6. Kayaking on Hanalei River 7. Boat cruise along the NaPali 8. Dolphin below Princeville 9. Kalalau Lookout 10. Sunday Polo at Anini. 11. Soaring by helicopter over the Grand Canyon of the Pacific and near waterfalls hundreds of feet high offers an opportunity to see the sights of Kauai from an angle that puts everything in a different perspective.

Princeville is all about golf. With 54 of the finest holes in the islands, it is truly a golfers paradise. The Makai course with 27 holes called the Lakes, Ocean and Woods, designed by Robert Trent Jones Jr. offers a championship layout with challenges for all levels of players. The Prince Course is considered Hawaii's most challenging and most beautiful.

Opposite Page

1. Morning golf on the Prince Golf Course 2.Horseback riding above Princeville 3. Old Hanalei Town 4. Diving at Anini Beach 5. Princeville Health Club 6. Couryard at Princeville Center.

There has always been something special about the northern shore of Kauai. From the days of the early Hawaiians who found the naturally fertile land a hospitable place to settle, to the missionaries who immediately recognized it as a place of great spirituality. The land has always been a source of peace, a contented way of life, and a home for all who have settled or visited.

Hanalei On The North Shore Of Kauai Is The Birthplace Of Hawaii

The historical significance of Princeville and the north shore cannot be overstated. The entire chain of Hawaiian islands was born of the eruption at Kauai's Kilauea volcano. The remains of the original volcano can be found at Crater Hill across Anini Bay and a short distance from Princeville.

Run-off from the verdant mountains surrounding Hanalei created the rich volcanic soil found on the Princeville plateau. It nourishes the lush hillsides, the flowering meadows, thickly overgrown ravines and the rich Hanalei Valley river bottom, where taro is grown today as it has been for generations.

Kauai's North Shore Is Blessed With Extraordinary Natural Beauty

Hanalei, the north shore's oldest community, is a true reflection of the island's aloha spirit. The historic buildings and sites that dot the lush rural landscape surrounding famed Hanalei Bay have changed little from the days when the missionaries landed.

There is a distinct sense of tranquility here. The towering mountains, the cascading waterfalls, the scent of ginger and plumeria, and the soft sea breezes make Hanalei a special place to relax and reflect.

The Centerpiece Of the North Shore Is The Masterplanned Community Of Princeville

In contrast to the slow pace of Hanalei, the master-planned destination resort community of Princeville sits on a grassy plateau above Hanalei Bay.

Set on a plateau above historic Hanalei, Princeville stretches along five and one half miles of Kauai's spectacular north shore.

Luxurious resorts, a variety of shopping, distinctive restaurants, lovely residential neighborhoods and world-class recreational facilities make Princeville one of Hawaii's most desirable vacation areas.

The Princeville Community Is Centered Around World Class Golf

Princeville's development began with the commissioning of Robert Trent Jones, Jr. to design a golf course that would unite the challenge of golf with a paradise setting. The original course, designed in 1969, is actually three distinct nine hole courses: the Ocean, the Woods, and the Lakes. In 1987 the Prince course added another eighteen holes for a total of fifty-four.

Jones' design philosophy is to respect and use the existing terrain to its best advantage. The many honors paid to Princeville affirm his success. The Makai and Prince Courses are rated among the top 100 in America. The Prince is considered one of Hawaii's most challenging courses. It is unquestionably one of most beautiful.

Respect For The Existing Environment Is The Criteria Of Princeville

Strict architectural design covenants guarantee the area's natural preservation. Emphasis has been placed on the use of native materials and the harmonious placement of structures within the existing terrain. Both landscape and building plans must be approved by a design committee. Locals like to say that it is written in the law that no building shall be taller than the tallest palm tree.

Although Princeville actually covers 11,000 acres, the developed portion barely occupies 1000 acres. This assures that generous amounts of open space are maintained. This is further enhanced by the wide golf course fairways that serve as massive greenbelt corridors.

Princeville provides residents and vacationers modern conveniences, luxurious living, and premier recreation without sacrificing the traditional values and inherent tranquillity of a rural community.

Ka'Eo Kai is as fresh as the day the resort opened nearly twenty years ago, because of PAHIO's dedication to responsible development and management. That same care and concern is evident in the quality construction of all PAHIO resorts. Over eighteen years of building experience in Hawaii has taught the developer that the islands' tropical climate has little respect for normal construction techniques, or inappropriate materials and designs used by mainland developers. On Kauai building with respect for the climate, as well as the land, requires an understanding that comes only with experience.

Since the dawning of man, every generation has shared a common desire to discover paradise, a place of perfection . . . a place where humanity and nature can co-exist in harmony. It is a quest that has proven so elusive that generations have turned to fantasy to temper their yearning to discover paradise. Ancient legends told of such a place in the land of Hanalei . . . where the mystical point at Bali Hai stands as the gateway to the jagged NaPali coast on the northern shore of Kauai. Its tropical beauty and serenity were said to rival that of the most exquisite visions. This is the vacation paradise of PAHIO. A place where the extraordinary splendor of Kauai combines with man made elegance.

PAHIO's Family Of Ownership Resorts Reflects Timeless Beauty Of Kauai

As Hawaii's most experienced vacation ownership company, PAHIO has an established reputation for building and managing only the finest resorts. Our four island resorts offer a unique combination of amenities. Each has a special ambience, yet maintains PAHIO's Spirit Of Excellence.

Many of our guests and owners prefer our spacious flagship resort, PAHIO at Ka'Eo Kai, because the 2200 square-foot two bedroom floor plan is perfect for the family.

Others rave about the breathtaking panoramic ocean view of the opulent Shearwater. Or the exclusive garden environment and sand bottom pool sculpted into the landscape of Bali Hai Villas.

Some prefer the convenient island location of Kauai Beach Villas, a beach front resort where they enjoy casual walks on the beach.

But every owner and guest appreciates our warm Aloha and the attention given to their vacation happiness.

The resorts of PAHIO occupy one of the world's most spectacular locations, so it was only appropriate that the resorts be as enduring as the land on which they are built. PAHIO has always been sensitive to the beauty of this special island and the heritage of its people. This strong appreciation, born out of respect for Kauai, dictates that PAHIO structures fit within the environment, that the design of buildings reflect the traditions of the island, and the resort creates as little impact on the natural surroundings as possible.

We Set The Standard For Vacation Ownership In Hawaii

Our one and two bedroom vacation villas define Gold Crown elegance. Every detail is considered. Designer furnishings and tropical accents reinforce the easy island lifestyle. Living rooms have entertainment centers. The gourmet kitchens feature convenient decorator appliances. Bathrooms are large . . . many have private spas. And, lanais offer spectacular views, yet provide privacy.

The warm aloha of PAHIO embraces every owner and guest from the moment they check-in.

Every PAHIO property reflects the sense of responsibility that is the company's trademark and demonstrates it's commitment to creating properties with unsurpassed style and quality.

Regard For The Vacation Lifestyle And Thoughtful Planning Make PAHIO Resorts Eminently Livable

A stroll through the lushly landscaped grounds of any PAHIO property confirms the company's emphasis on maintaining the ambiance of Kauai. Walkways wind through gardens of flowering plants and fruit bearing trees. The positioning of buildings and the equally careful placement of the windows in each villa invites the aroma of plumeria, hibiscus and ginger into the living spaces. Every room has a view. Every room gets the benefit of the sea breeze. Every room is a reflection of the traditional Hawaiian lifestyle.

Established in the early 1980's, PAHIO is recognized internationally for superior standards of quality and service. Over the years, our owners and guests have come to expect the finest vacation possible . . . A quality Aloha experience that only PAHIO Resorts can offer.

PAHIO's standardized cleaning procedures and checking system guarantee consistent quality year after year.

Swimming or simply enjoying an afternoon at Queen's Bath is a truly unique Princeville experience.

PAHIO's experienced vacation counselors are longtime Kauai residents and are trained to help you plan a memorable vacation. PAHIO offers reduced rates on many island activities.

The legacy of Bali Hai Villas is hidden in a ravine that runs through the resort. Nestled into the cliff where the ravine plunges to the ocean you will find a natural seawater spa formed by a series of rocks that are filled with the tide and warmed by the sun. Known as the Queen's Bath, the pools served as a soothing respite for ancient Hawaiian royalty on holiday in Hanalei. The pools that were once exclusively reserved for queens are now part of the enchantment of Bali Hai Villas. The resort is privileged to have many such treasures.

The Site Of Bali Hai Villas Holds An Important Piece Of Kauai's Heritage

There isn't a more important site for a resort within Princeville than the one chosen for Bali Hai Villas. Located on the high ground near the Makai Clubhouse, the resort is a focal point of the community. As such, the resort's landscaping and architecture had to be sensitive to Princeville's stringent design criteria.

Bali Hai Villas is destined to become more than the standard of excellence for vacation ownership . . . it will be the standard by which Princeville is judged as a premier vacation destination.

PAHIO Reshaped The Land To Create A Paradise Within A Paradise

PAHIO's desire to create an exclusive vacation environment is especially apparent at the pool and recreation area. The land was reshaped to create a private alcove for the pool and spa while reducing the visual impact of the common area for nearby villas. The contoured habitat begins with a series of sculpted waterfalls that cascade over landscaped boulders, eventually flowing into the large fresh water swimming pool. The pool's unique design is highlighted with an underwater reef that forms a separate kiddies pool and a sand bottom that continues beyond the pool to make a beach for building sand castles and playing volleyball.

A large cabana for entertaining completes the recreation habitat. It features restrooms, cooking grills, tables, and a kitchen.

Bali Hai Villas Is A Sanctuary From The Rigors of Reality

For those who have always desired a tropical island retreat, PAHIO has thoughtfully created the perfect vacation hideaway. Residences have a wonderful view of the mountains, the ocean, and the tropical setting created within the resort.

No compromises were allowed, each villa is built with quality craftsmanship and offers superb finishing touches. Interior appointments and colorful designer accents have a Balinese influence. Gentle island breezes waft through each villa from ceiling fans. And, spacious lanais provide an inviting outdoor complement to the airy floor plan and gracious interior design.

Hawaii's Casual Aloha Attitude Influenced The Interior Design

In the living room, the soft pillow comfort of a stylish rattan couch and an attractive glass top coffee table provide a comfortable focus. The six-foot sofa unfolds to become a queen-size bed for extra guests or large families. Matching chairs provide a personal touch for

conversations or reading. An entertainment center with a television, am-fm stereo system, and a video playback completes the living room ensemble. The designer unit is built into the wall to create the appearance of a foyer as well as hide the electronics.

King-sized beds, tropical decorations and gentle relaxing colors highlight the spacious bedrooms. Cooled by the ceiling fans and soft romantic tradewinds, you will find the bedrooms are a perfect retreat after the day's adventures. Each bedroom has a remote controlled cable television, rattan night stands, and dresser.

Over the years, PAHIO research studies have indicated that owners and guests appreciate spacious and well designed bathrooms. Bali Hai Villas' bathrooms feature the finest fixtures available. Hair dryers in the bathrooms, and a large tub with a spa in the master bath complete the relaxing ambience.

The kitchen facility is a great convenience for the vacationing family. It can stretch the vacation dollar while providing a welcome break from the restaurant routine. Special appointments include a toaster, blender, microwave, range with oven, ice-making refrigerator . . . and all the pots, pans, and utensils necessary to prepare a gourmet meal.

Experience The Magic Of A PAHIO Vacation

Simply watching and enjoying Kauai's beautiful north shore from the lanai becomes a favorite pastime for many resort guests. For this reason PAHIO selected outdoor furniture for Bali Hai Villas that is extremely comfortable, as well as durable enough to resist the elements.

Below the Shearwater, the crescent Anini Bay is sheltered by a natural coral reef a few hundred yards off-shore. Within its protective walls the sea is calm, and local fishermen harvest lobsters while visitors safely snorkel and windsurf the warm, gentle water. Beyond the edge of the reef, where the waves break, dolphins play and humpback whales pass on their annual migration . . . nothing in your imagination can match the stunning spectacle of one of these gentle giants rising out of the sea in joyful tribute to the land.

Days And Nights Weave Their Separate Magic On The Shearwater

Perched on a bluff above pristine Anini Beach, the Shearwater commands a breathtaking panoramic view stretching from a point beyond Hanalei's romantic Bali Hai in the west to the historic Kilauea Crater and Lighthouse in the east . . . nothing interrupts the view except the occasional, graceful Albatross or Shearwater bird floating on a thermal updraft near the cliff's edge, and near your lanai. It is truly a location that defines the term "tropical paradise."

The Shearwater Sits At The Edge Of Paradise

The splendor of the bay is ever changing. In the morning, a mist rises from the water, draping the horizon in a surreal veil as the golden light of sunrise sparkles off the dew splashed cliffs. By noon the sun burns-off the mist and turns the color of the bay to a turquoise, revealing the fruitful coral beds. In the afternoon the core of the island becomes exposed in the warmth of the sun only to disappear into a shroud of myth and mystery as the evening mist recaptures the mountain's summit.

As the sun gives way to twilight, the sky shines in a spectacular display that can be experienced nowhere else but here. The setting sun seems to accentuate the sound of the sea breaking against the reef. As the tropical moon rises over Hanalei Bay and the trade winds begin to whisper the romantic songs of this paradise, time loses its meaning and you are swept away by the refreshing sounds of silence.

As a PAHIO resort, the Shearwater captures the essence of Kauai. A stroll through the lushly landscaped grounds confirms the resort's emphasis on maintaining the ambience of the Garden Island. The careful orientation of each villa takes full advantage of the cool tradewinds and the spectacular views offered by the site.

The Shearwater is a sanctuary from the rigors of reality. It is a place of peace and lush tropical tranquility. This is the rarest of locations for a resort . . . perhaps the last of it's kind in Hawaii, and the world.

Enjoy The Seclusion Of The Shearwater

This exclusive collection of truly opulent two bedroom, two bath villas defines Gold Crown elegance. Every luxurious convenience has been considered to make vacationing in a Shearwater villa as comfortable as possible.

The living room includes an entertainment center with a television, stereo and video system, and stylish overstuffed rattan furnishings including a queen size sofa sleeper.

The gourmet kitchen has solid wood cabinets, decorator appliances and convenient amenities like a Jenn-Air cooktop range and oven, dishwasher and a microwave. An ensemble of cooking utensils and dining service for eight make vacation cooking and entertainment a delight.

Cooled by the ceiling fans and soft romantic tradewinds, you will find the bedroom is a perfect retreat from your day's adventures. Tropical decorations compliment the gentle colors of these relaxing rooms. Each bedroom has a king-sized bed, remote controlled cable TV, rattan night stands, dressers, and a combination writing desk with a mirrored vanity. And both of the master-size bathrooms have a large tub with a soothing spa, and a separate shower.

Other amenities include an accessorized wet bar, decorator ceiling fans, washer and dryer, and a full compliment of lanai furniture.

The Shearwater, like the island of Kauai, is a world apart. Removed from the rush of the mainland by both location and time, it is a place where you can happily sit for hours watching the valleys for the birth of a new rainbow . . . the ocean for the breaching of a whale . . . or the blue Hawaiian sky for a glimpse of the endangered Shearwater bird soaring on an ocean updraft.

PAHIO schedules weekly luaus and activites where owners and guests mingle and learn to dance the hula.

If Kauai Beach Villas offered nothing more than its exotic environment, it would be a vacation paradise . . . but of course, the resort provides much more to make your vacations memorable, year after year.

Every PAHIO villa has a washer and dryer.

It is said that the land surrounding Kauai Beach Villas was a spiritual retreat for ancient warriors. Today the magic and serenity of the site has been captured by the resort's beautifully landscaped lagoon . . . a man-made environment so natural that it has been adopted by wildlife. Here friendly swans entertain guests while watching over the lagoon. Young families share a peaceful afternoon as the children play with tame ducks. And couples enjoy an evening stroll along the path winding between flower gardens or a restful moment on a bridge.

Kauai Beach Villas Offers A Perfect Blend Of Style, Location And Value

The best of Kauai is within minutes of your retreat. The resort's location on Kauai's eastern shore, known as the Coconut Coast, affords convenient access to the landmarks and activities that have made the island such a popular vacation destination.

Walk along the beach, or simply enjoy a lazy day in the sun. The ocean is just footsteps from the resort. Within minutes there are challenging golf courses on par with the best in the world. If shopping is on your agenda, rustic nearby towns offer unique island style boutiques, shopping centers, bookstores and pubs rich in character. Nearby, four star restaurants and local eateries famous for serving abundant portions of good food await exploration.

Just minutes from Kauai's main airport, the resort is seemingly in the center of the island. North of the resort, you'll discover the Fern Groto on Wailua River, Hanalei Valley with its taro fields, romantic Hanalei Bay, the historic Waioli Mission, and the breathtaking NaPali Coast.

To the south, you will sunbath and snorkel in sunny Poipu, visit rustic Koloa Town, explore the pine forests at Kokee and photograph the Grand Canyon of the Pacific.

The Resort Offers Fine Dining And Great Entertainment

Dine in a fine restaurant, dance or chuckle at a comedian performing in the night club. It's all yours at Kauai Beach Villas. Because the resort shares its prime ocean front location with a major hotel, PAHIO owners have the best of two worlds. . . the convenience, economy and privacy of a condominium resort plus the glamour and entertainment that is available only at a premier resort hotel with restaurants and night club entertainment.

The Kauai Beach Villas Is A Resort For The Entire Family

Long considered one of the island's best kept secrets, Kauai Beach Villas is equally rich in style and amenities. Tennis enthusiasts will appreciate the beautiful lighted courts.

Kauai Beach Villas has two beautiful swimming pools. The private pool and intimate spa at the resort's pavilion is perfect for a quick refreshing dip after the day's activities.

With a waterfall, wading pool and spa, the hotel's large sculptured pool is a grand family playground. Children enjoy the shallow wadding section. Adults appreciate the exercise afforded in the remaining portion of the pool. The entire family can relax in the spa. And there are plenty of deck chairs available around the pool for sunbathing.

Next to the pool there is a snack stand that serves beverages, snacks and delightful light lunches.

Interior Appointments Feature Tropical Elegance

Hawaii's casual aloha attitude influenced the interior decor of the Kauai Beach Villas. Throughout, white washed rattan, cabinets and taupe carpeting contrast the bold flower patterns and rich hues selected for the upholstery, wall treatments and bedding.

In the living room rattan conversation chairs, an overstuffed chaise lounge with an ottoman, and a matching sofa that unfolds to become a queen bed for large families create a tropical theme.

An entertainment center with a 27-inch television, am-fm stereo system, and a video playback completes the living room ensemble

The master bedroom has a king bed and a private bath, while second bedrooms have either a queen or twin beds.

An Informal Kitchen Accommodates Vacation Dining With Style

PAHIO realizes the importance of a well planned kitchen for the vacationing family and has given special attention to this often overlooked interior amenity.

Easy preparation and quick clean-up were a kitchen priority. There is plenty of wrap-around counter space permitting easy access to the built-in range, double sink, dish washer and refrigerator. The low profile counter works nicely for casual meals and conversations with the cook. Additional amenities include a toaster, ice making refrigerator, blender, microwave, . . . and all the necessary pots, pans, and utensils.

Lumahai Beach
Hanalei Bay
2
3
4
Princeville
Haena
NaPali Coast
Old Hanalei Town
Kalalau Lookout
Waimea Canyon
KAUAI
Kapaa
Opaekaa Falls
Barking Sands
1
Lihue
Waimea
Koloa
Poipu